THE POLLEN GRAIN DRAWINGS OF DOROTHY HODGES

Taken from the Pollen Loads of the Honeybee

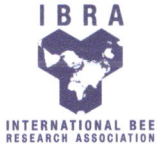

The Pollen Grain Drawings of Dorothy Hodges
Taken from the Pollen Loads of the Honeybee
ISBN: 978-1-913811-07-5

Published 2021 by:

IBRA, 1 Agincourt Street, Monmouth, NP25 3DZ (UK).
www.ibra.org.uk

Northern Bee Books, Scout Bottom Farm,
Mytholmroyd, Hebden Bridge HX7 5JS (UK)
www.northernbeebooks.co.uk

Back cover image: Piazza Barberini, Rome 1958. Copyright and use is granted by the Eva Crane Trust.

IBRA Proof-Editor: Stuart A. Roberts; design by: www.SiPat.co.uk

All rights reserved. No part of this publication may be reproduced, stored or transmitted in any form or by any means electronically or mechanically, by photocopying, recording, scanning or otherwise, without the permission of the copyright owners

© 2017, IBRA

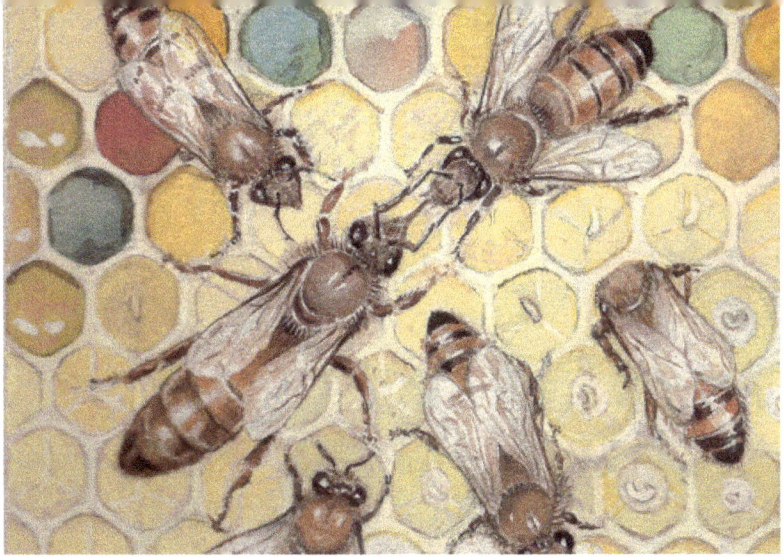

Dorothy Hodges was a trained artist with an artist's acute powers of observation as well as being a beekeeper. In 1946 she had the idea of making a colour chart of pollen loads. It took several years for her ideas to gestate but the glorious outcome was the publication, by the then Bee Research Association, of The Pollen Loads of the Honeybee in 1952.

Designed as a very practical guide for beekeepers, the importance of the book was immense and it has long since risen from being a humble textbook and guide to a much sought after collectors' item. Its rarity and importance mean that it is no longer easily obtainable and so difficult for the ordinary beekeeper to appreciate its contents. For this very reason IBRA has decided to reproduce Mrs Hodges's delicate drawings of pollen grains as a separate publication and in so doing hopefully make her work known to other generations of beekeepers.

Although the painstakingly produced colour charts of the original book still have their value it would.not be possible to reproduce them with sufficient accuracy to do justice to the original work. However, the drawings lend them-selves to reasonable reproduction. They are of outstanding artistic merit and offer the possibility of identifying the pollen forms which are most frequently collected by bees. For beginners these drawings will do good service as an introduction to the pollen analysis of honey.

The drawings need no explanation other than a name - the family group, the Latin scientific name and the common English name - thus making the book independent of language barriers. This means it can be appreciated in many countries where the original work was unknown or is now out of reach because of rarity and cost.

The cover is taken from Dorothy Hodges own watercolour painting that she suggested might adorn the dust jacket of the original publication. The artwork was not used and so this booklet allows it to be seen publicly for the first time in almost sixty years.

Finally, for the convenience of the reader, the actual pollen drawings retain the same page, numbers as the plates in the original book.

Richard Jones
Former Director, IBRA
October 2009

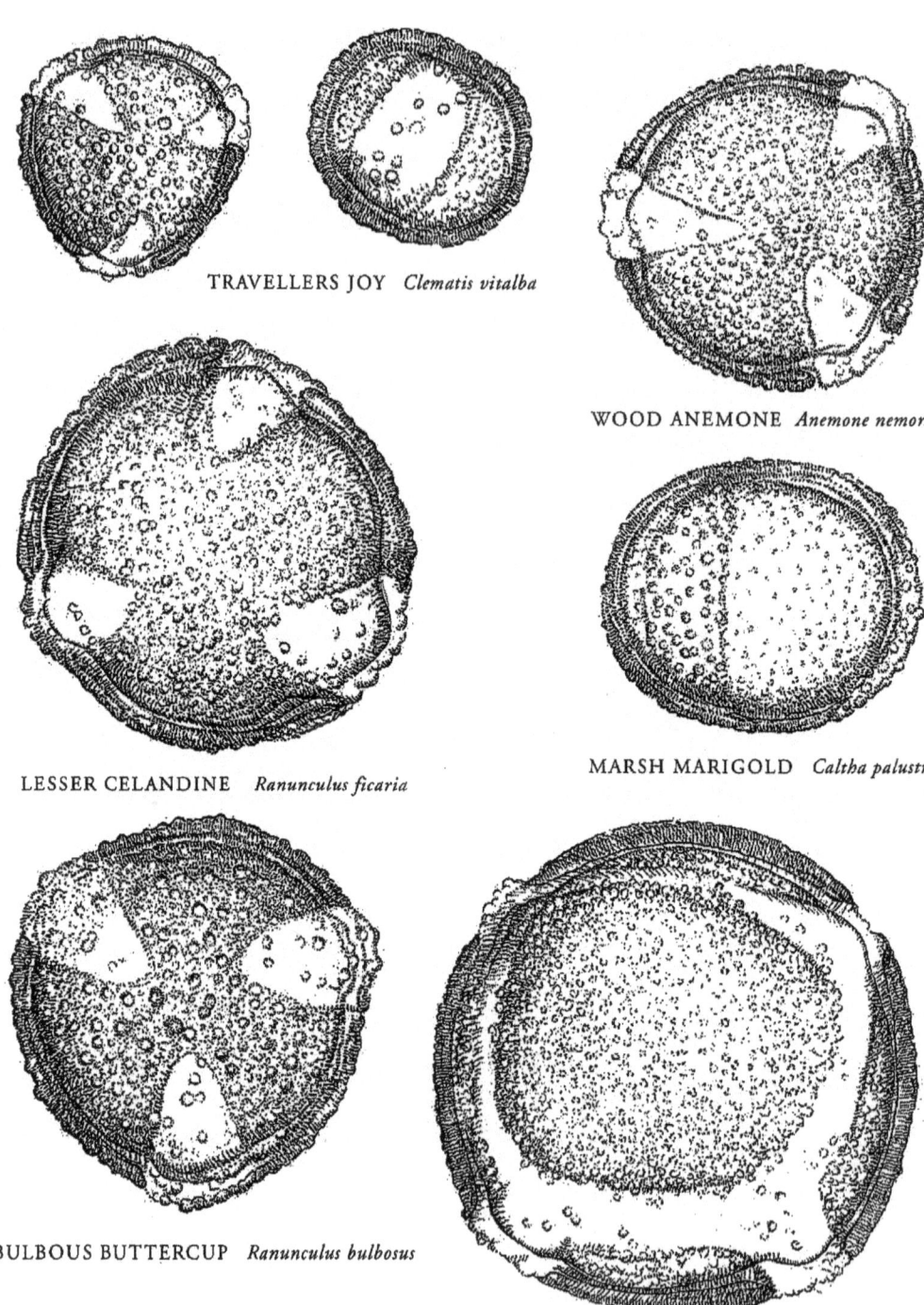

TRAVELLERS JOY *Clematis vitalba*

WOOD ANEMONE *Anemone nemorasa*

LESSER CELANDINE *Ranunculus ficaria*

MARSH MARIGOLD *Caltha palustris*

BULBOUS BUTTERCUP *Ranunculus bulbosus*

x 1500 d.

20 microns

BERBERIS *Mahonia aquifolia*

RANUNCLUACEAE   BERBERIDACEAE

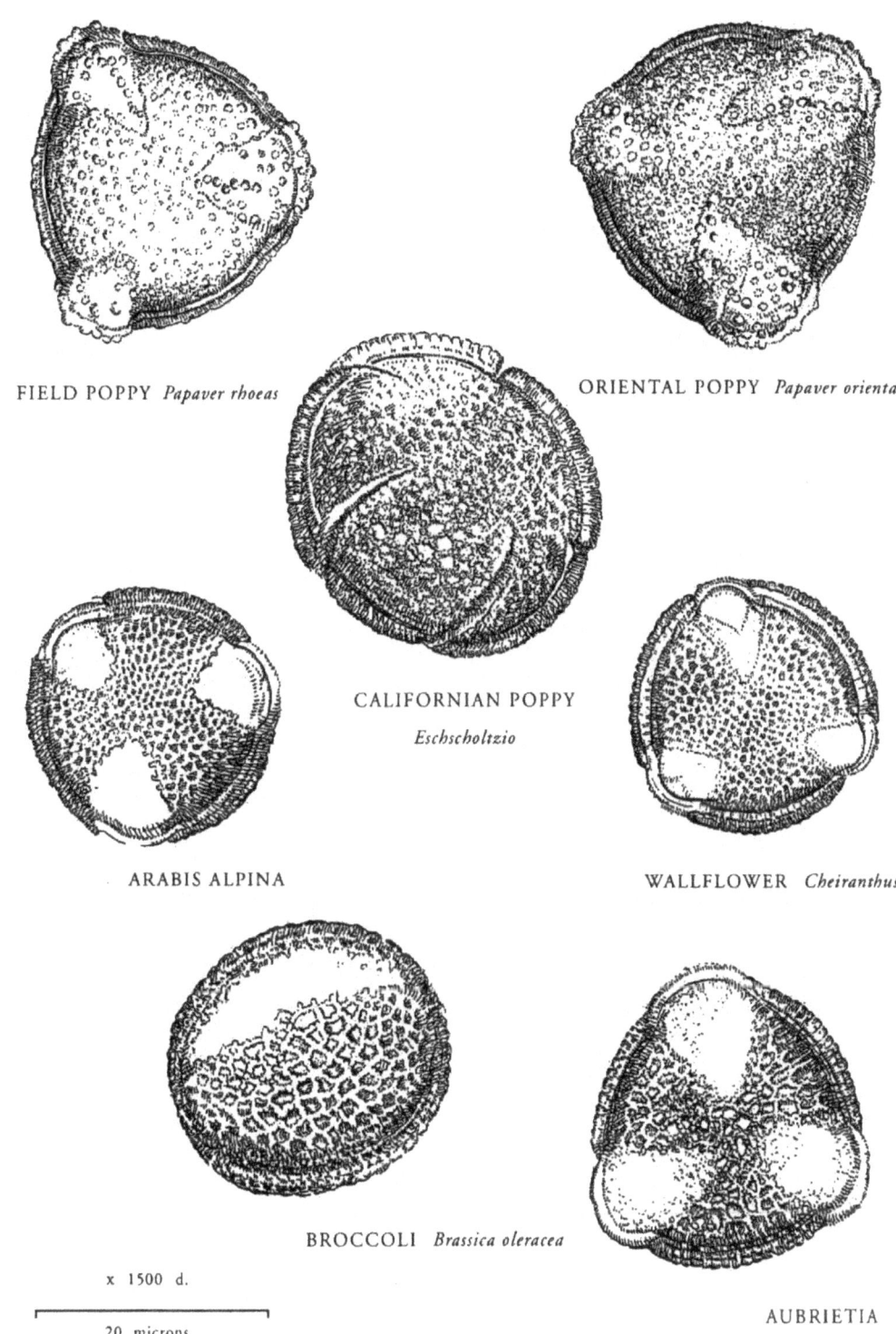

FIELD POPPY *Papaver rhoeas*

ORIENTAL POPPY *Papaver orientale*

CALIFORNIAN POPPY *Eschscholtzio*

ARABIS ALPINA

WALLFLOWER *Cheiranthus*

BROCCOLI *Brassica oleracea*

AUBRIETIA

x 1500 d.
20 microns

PAPAVERACEAE   CRUCIFERAE

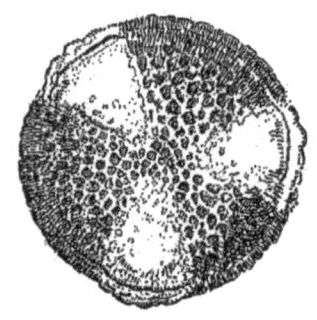
WHITE CHARLOCK *Raphanus raphanistrum*

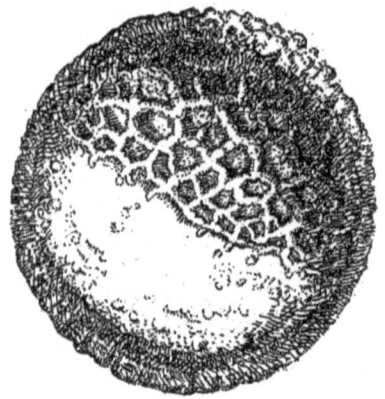
CHARLOCK *Sinapis arvensis*

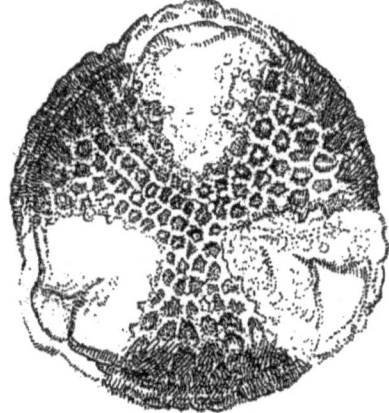
CHARLOCK *Sinapis arvensis*

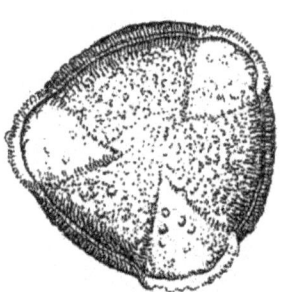
MIGNONETTE *Reseda odorata*

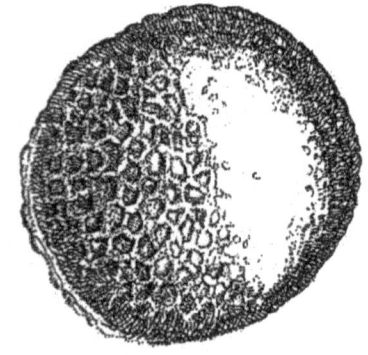
WHITE MUSTARD *Sinapis alba*

x 1500 d.

20 microns

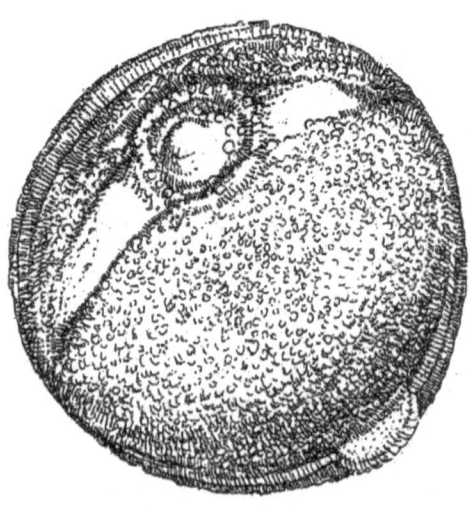
GARDEN ROCKROSE *Helianthemum*

CRUCIFERAE   RESEDACEAE   CISTACEAE

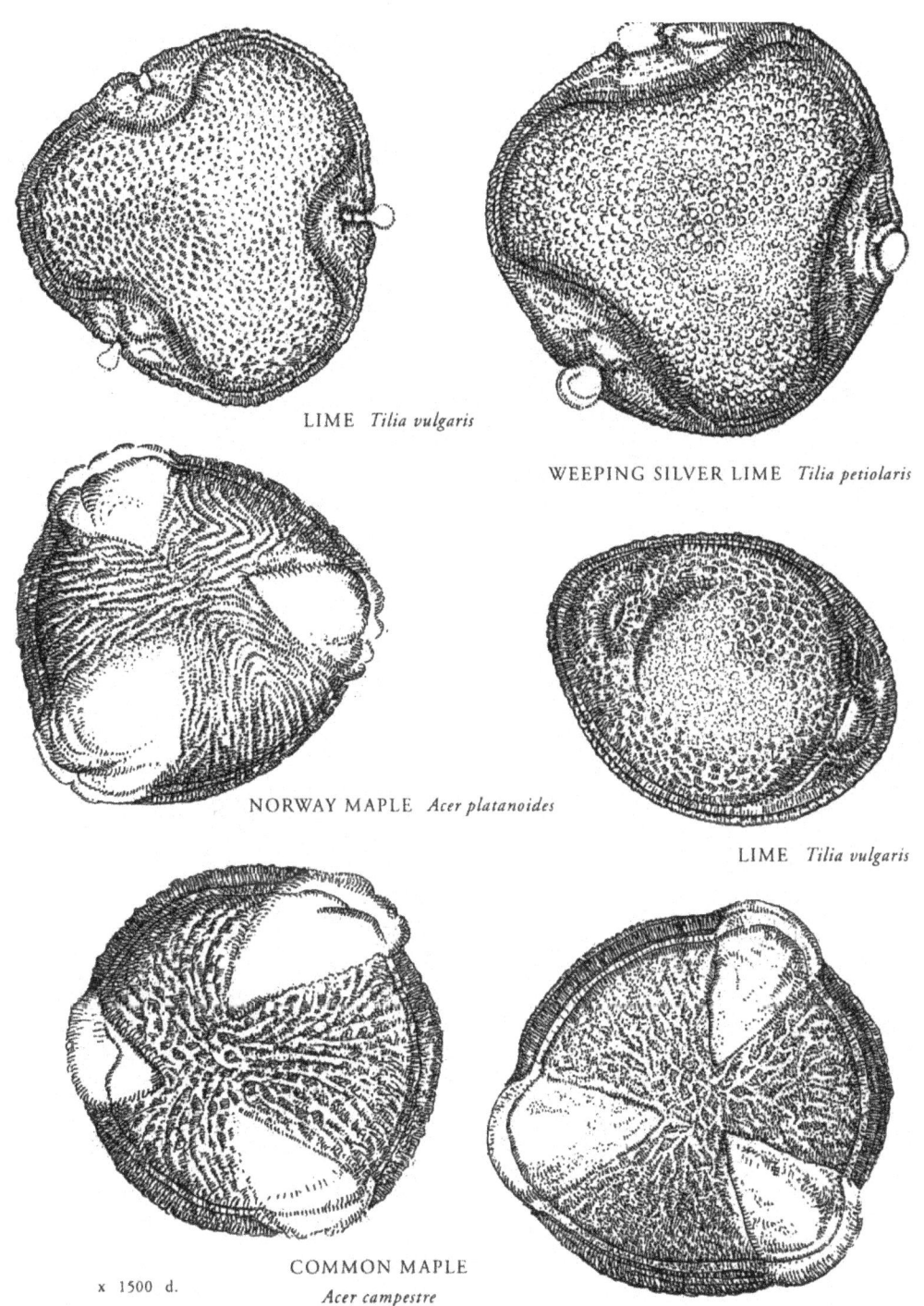

LIME *Tilia vulgaris*

WEEPING SILVER LIME *Tilia petiolaris*

NORWAY MAPLE *Acer platanoides*

LIME *Tilia vulgaris*

x 1500 d.

20 microns

COMMON MAPLE
Acer campestre

SYCAMORE *Acer pseudo-platanus*

TILIACEAE   ACERACEAE

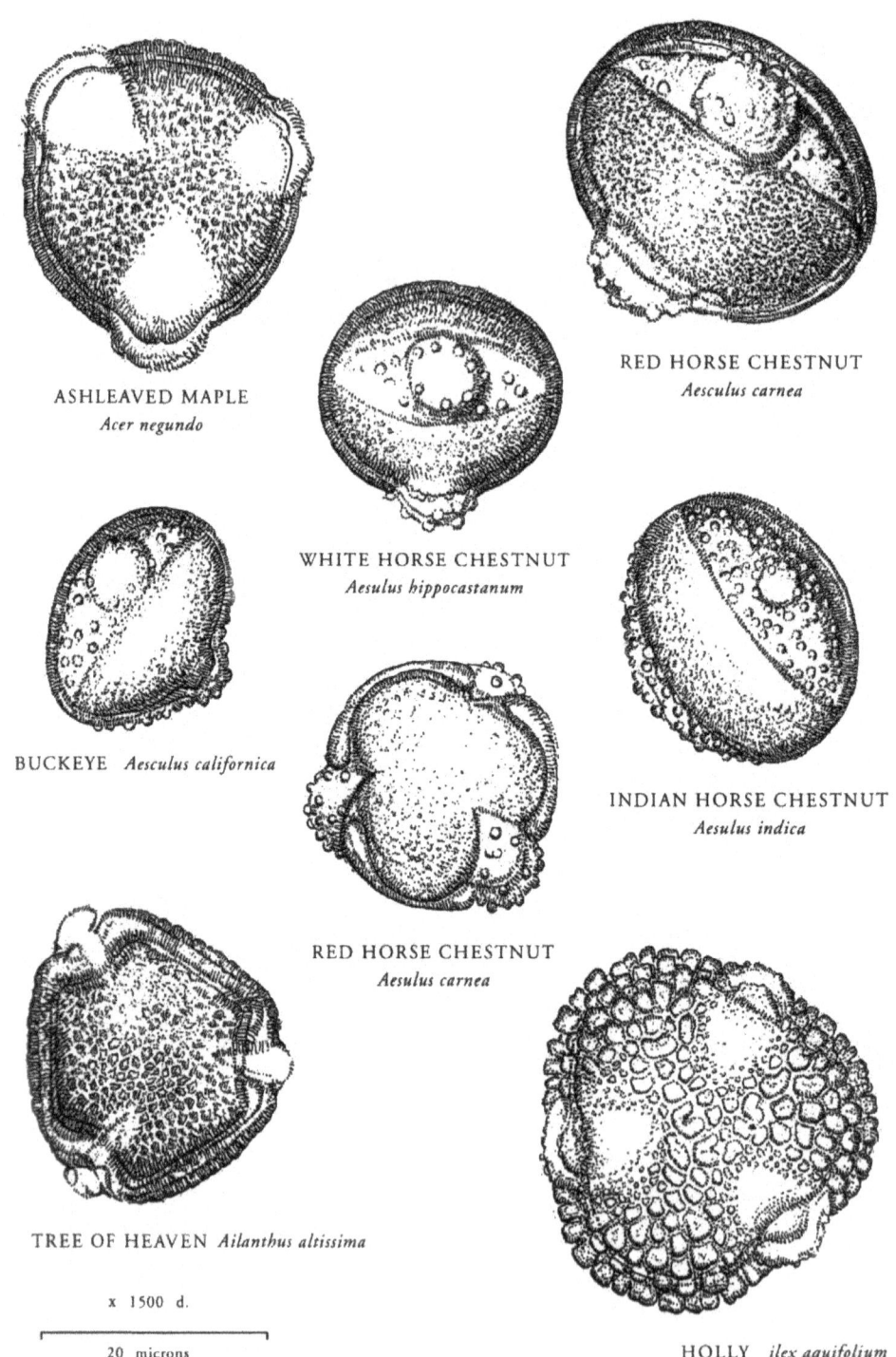

ASHLEAVED MAPLE
Acer negundo

RED HORSE CHESTNUT
Aesculus carnea

WHITE HORSE CHESTNUT
Aesulus hippocastanum

BUCKEYE *Aesculus californica*

INDIAN HORSE CHESTNUT
Aesulus indica

RED HORSE CHESTNUT
Aesulus carnea

TREE OF HEAVEN *Ailanthus altissima*

x 1500 d.

20 microns

HOLLY *ilex aquifolium*

ACERACEAE   HIPPOCASTANACEAE   AQUIFOLIACEAE   SIMARUBACEAE

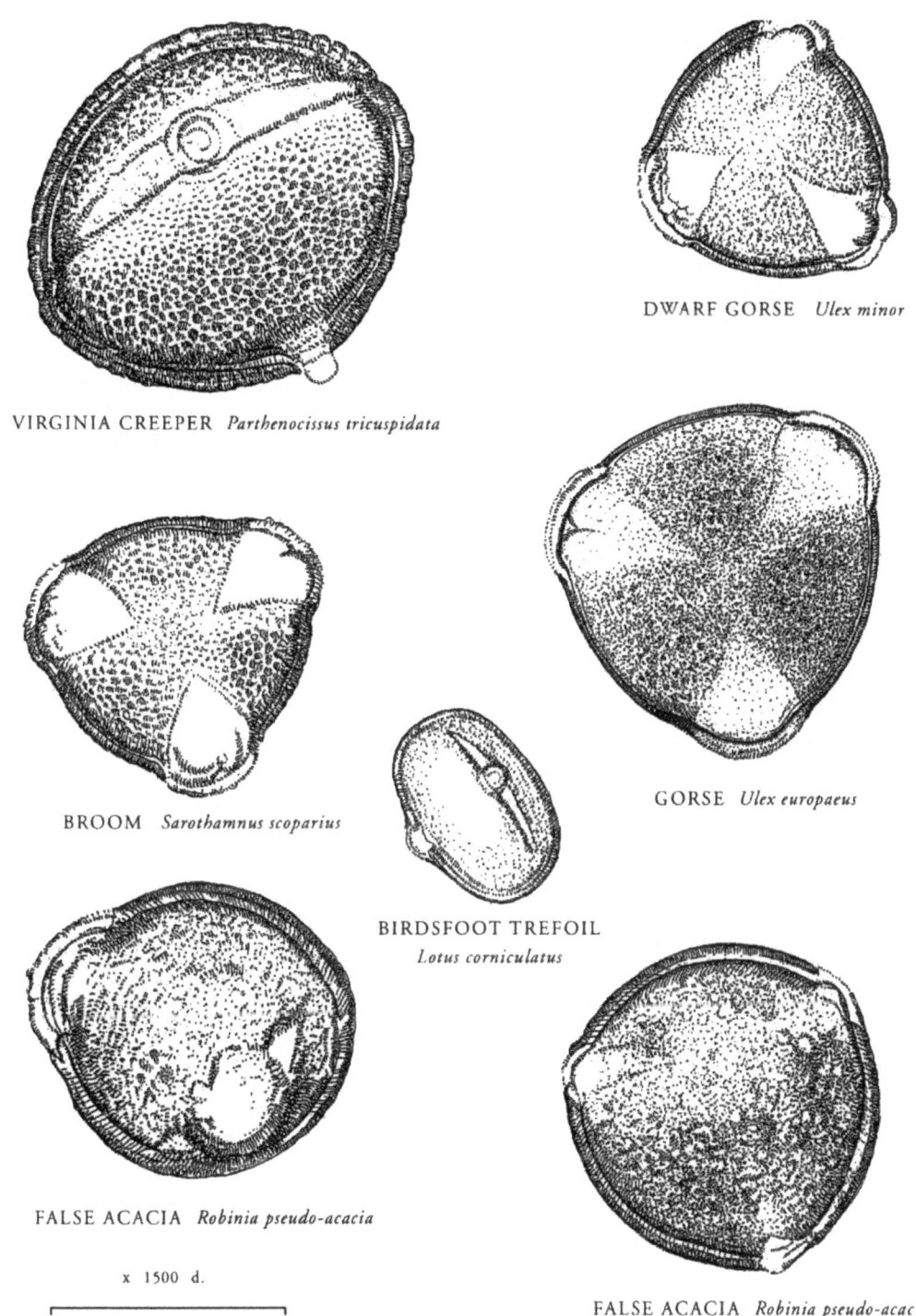

VIRGINIA CREEPER *Parthenocissus tricuspidata*

DWARF GORSE *Ulex minor*

BROOM *Sarothamnus scoparius*

BIRDSFOOT TREFOIL *Lotus corniculatus*

GORSE *Ulex europaeus*

FALSE ACACIA *Robinia pseudo-acacia*

x 1500 d.

20 microns

FALSE ACACIA *Robinia pseudo-acac*

RHAMNACEAE   LEGUMINOSAE

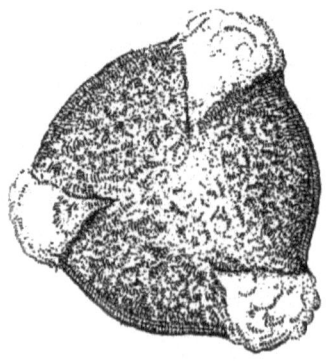
WHITE CLOVER *Trifolium repens*

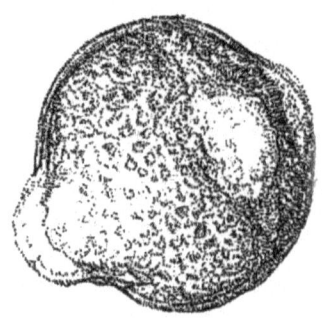
WHITE CLOVER *Trifolium repens*

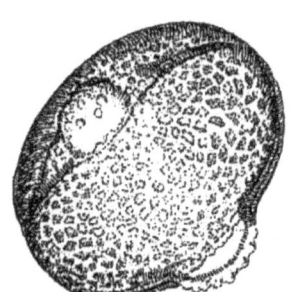
YELLOW MELILOT *Melilotus officinalis*

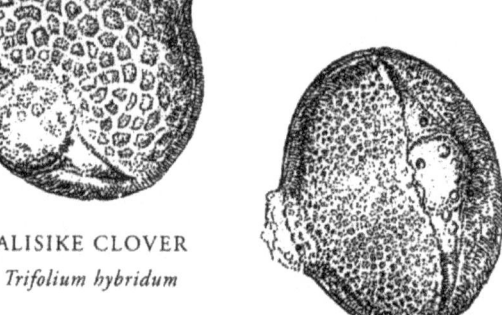
ALISIKE CLOVER
Trifolium hybridum

WHITE MELILOT *Melilotus alba*

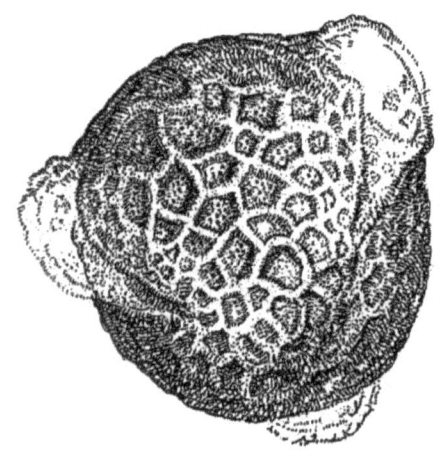

x 1500 d.

20 microns

RED CLOVER
Trifolium pratense

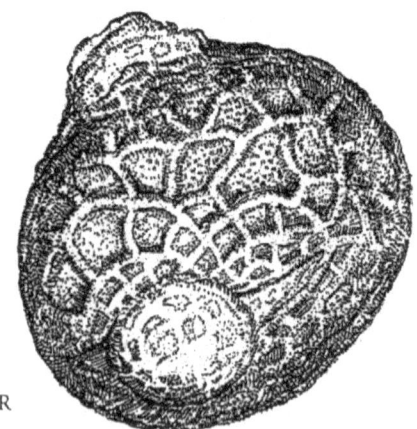
RED CLOVER *Trifolium pratense*

LEGUMINOSAE

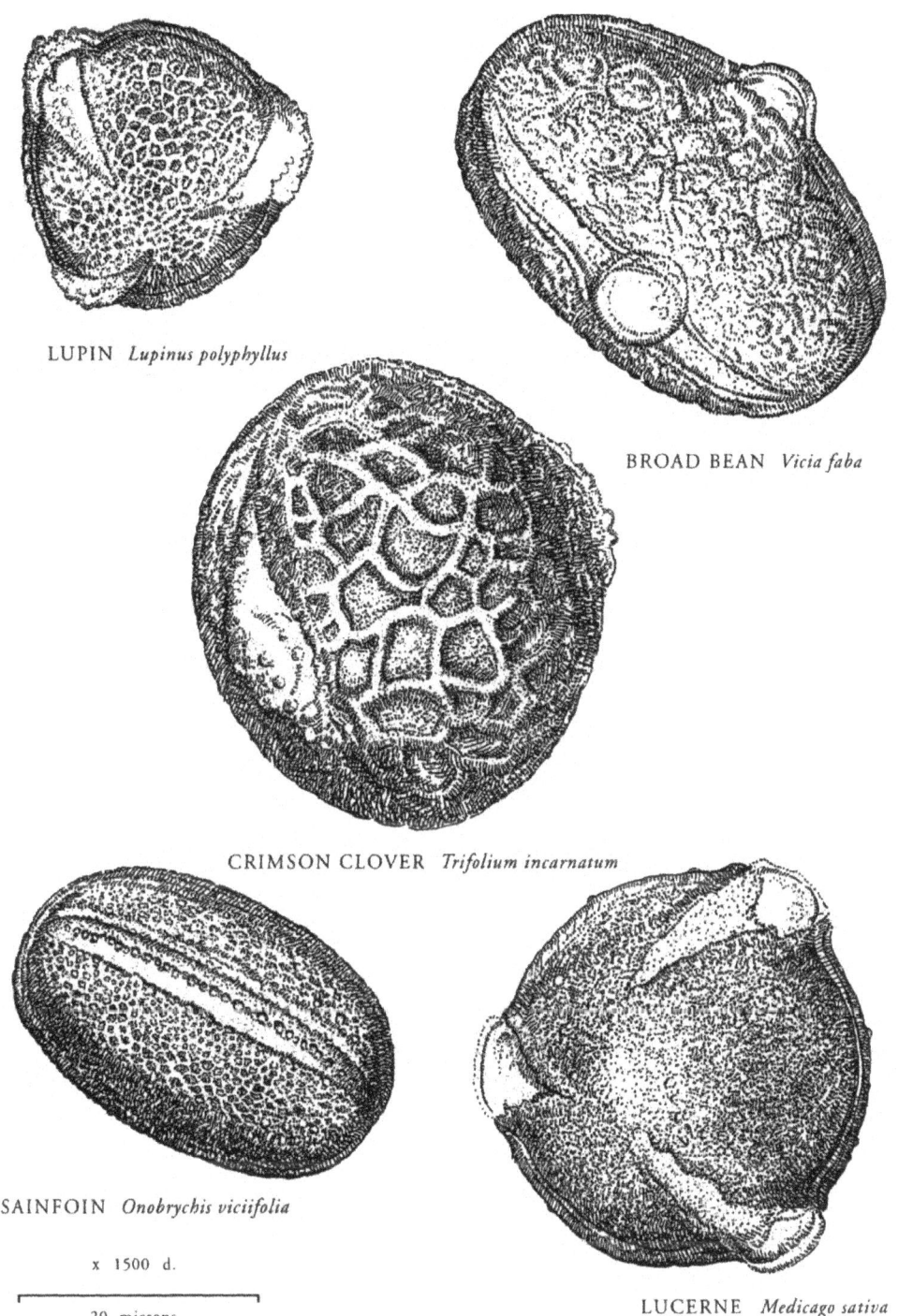

LUPIN *Lupinus polyphyllus*

BROAD BEAN *Vicia faba*

CRIMSON CLOVER *Trifolium incarnatum*

SAINFOIN *Onobrychis viciifolia*

x 1500 d.

20 microns

LUCERNE *Medicago sativa*

LEGUMINOSAE

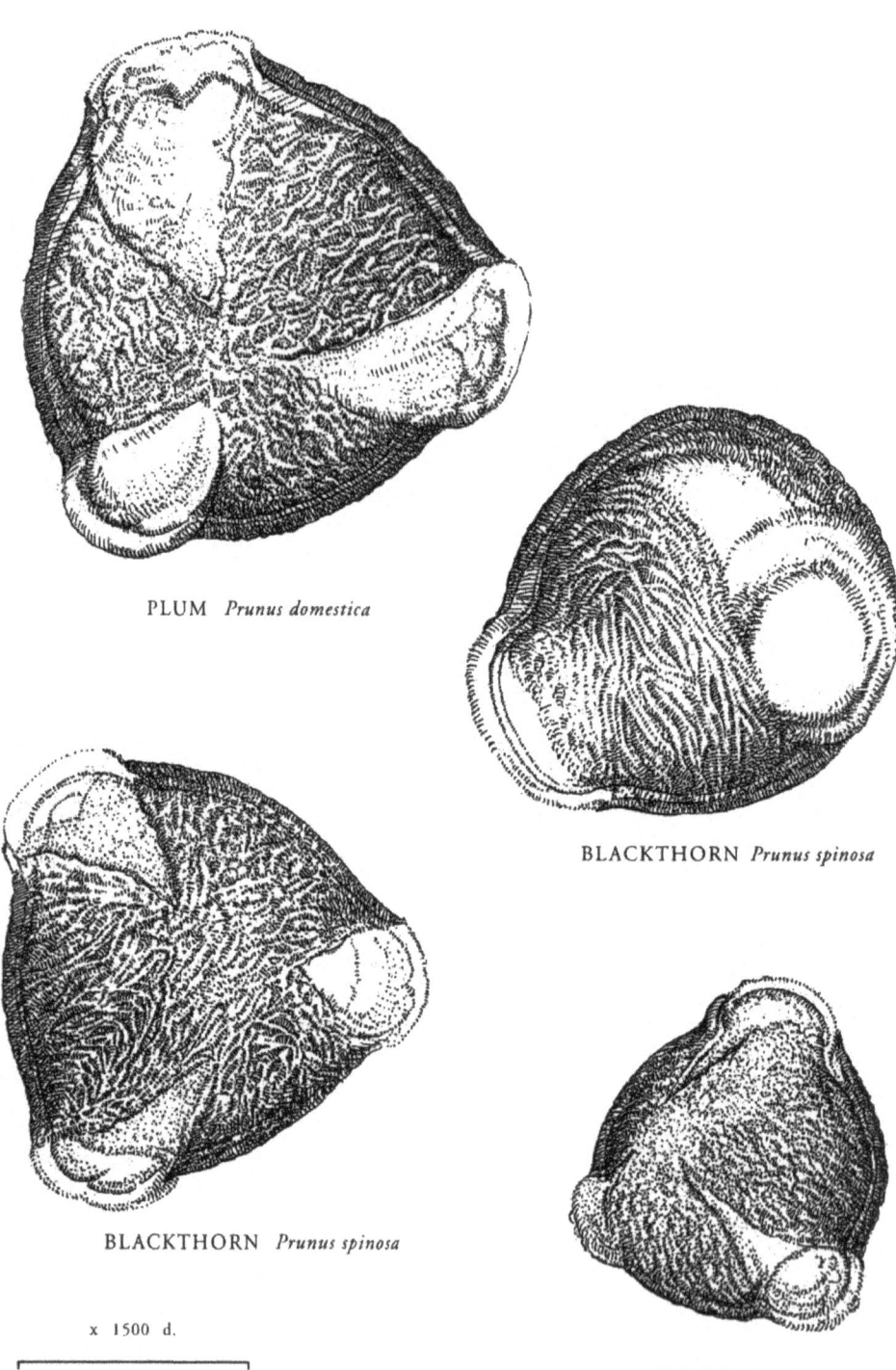

PLUM *Prunus domestica*

BLACKTHORN *Prunus spinosa*

BLACKTHORN *Prunus spinosa*

DOG ROSE *Rosa canina*

x 1500 d.

20 microns

ROSACEAE

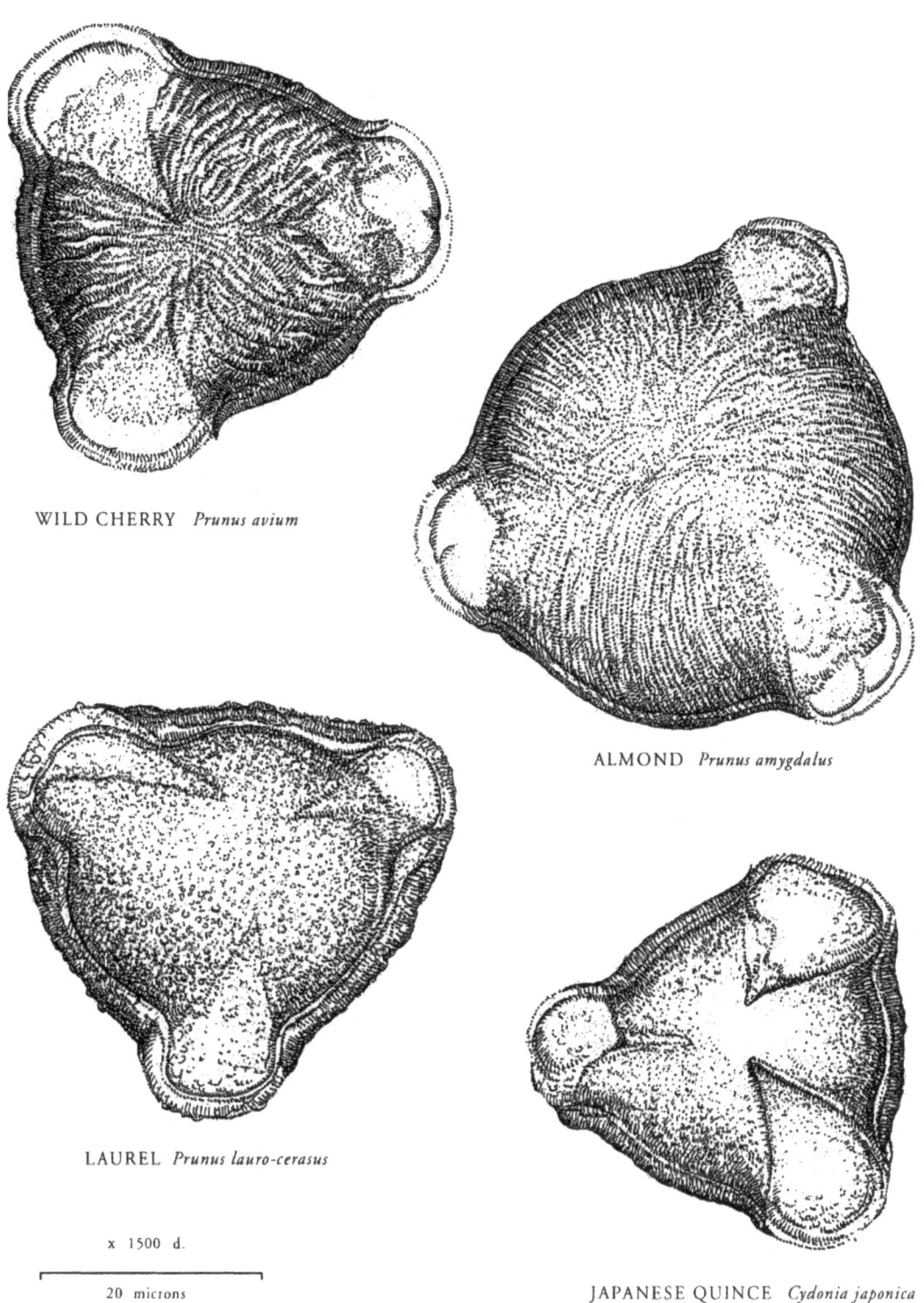

WILD CHERRY *Prunus avium*

ALMOND *Prunus amygdalus*

LAUREL *Prunus lauro-cerasus*

x 1500 d.

20 microns

JAPANESE QUINCE *Cydonia japonica*

ROSACEAE

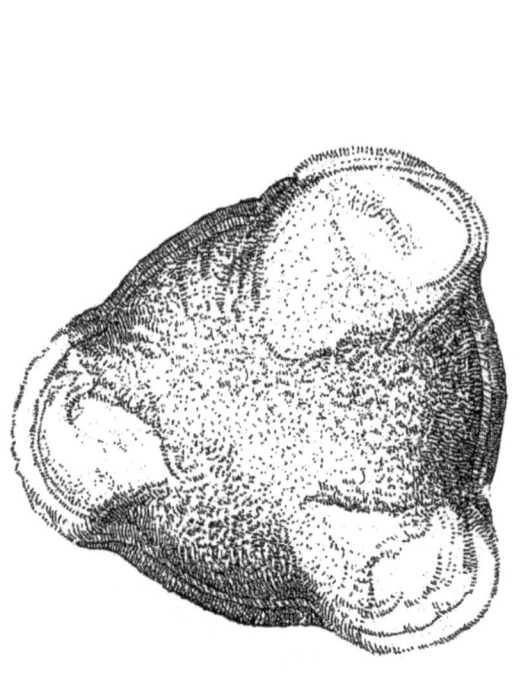

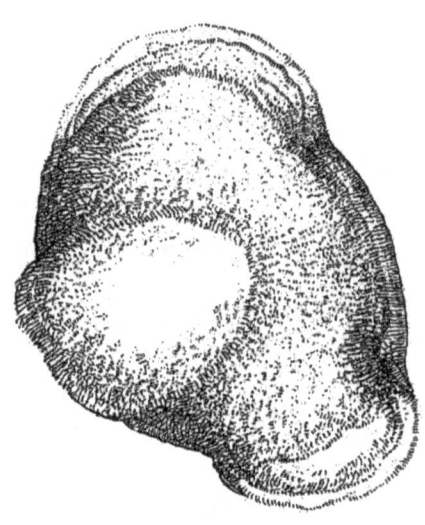

HAWTHORN *Crataegus monogyna*

HAWTHORN *Crataegus monogyna*

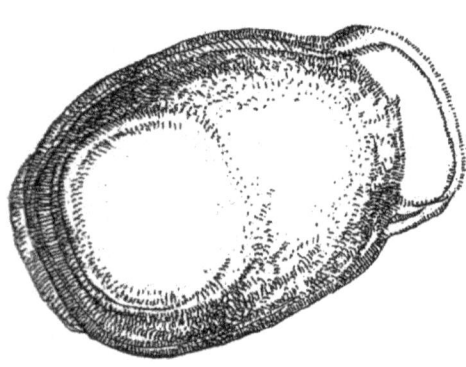

APPLE *Malus pumila*

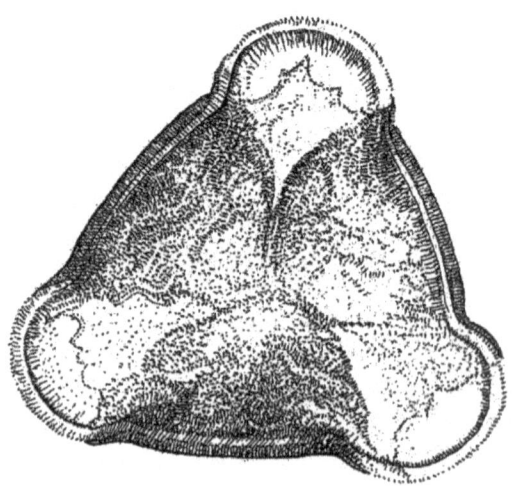

APPLE *Malus pumila*

x 1500 d.

20 microns

ROSACEAE

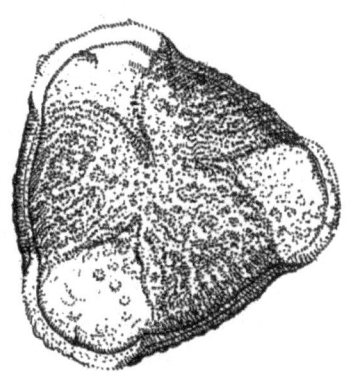
RASPBERRY *Rubus idaeus*

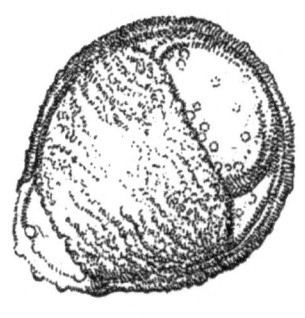
RASPBERRY *Rubus idaeus*

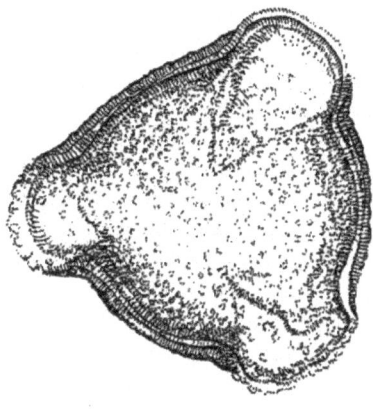
BLACKBERRY *Rubus fruticosus*

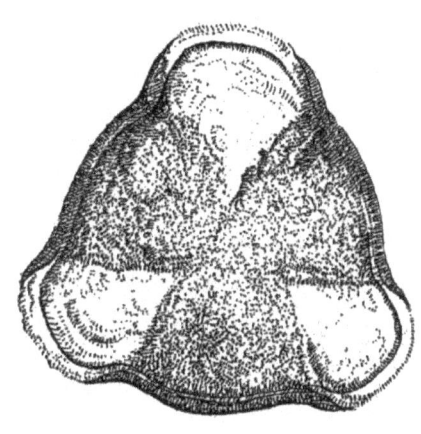
PEAR *Pyrus communis*

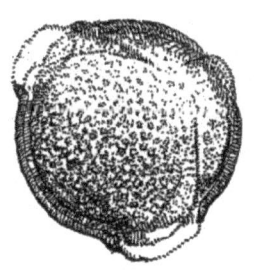
MEADOWSWEET *Filipendula ulmaria*

x 1500 d.

20 microns

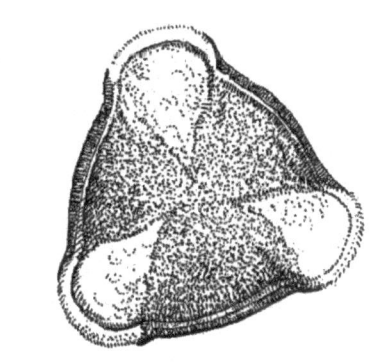
MOUNTAIN ASH *Sorbus aucuparia*

ROSACEAE

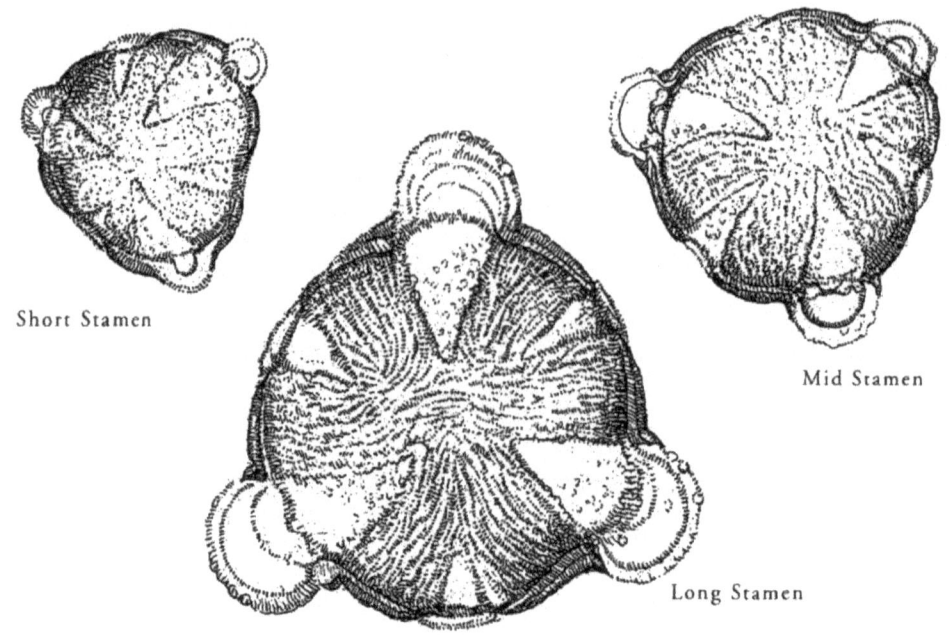

Short Stamen

Mid Stamen

Long Stamen

PURPLE LOOSESTRIFE *Lythrum salicaria*

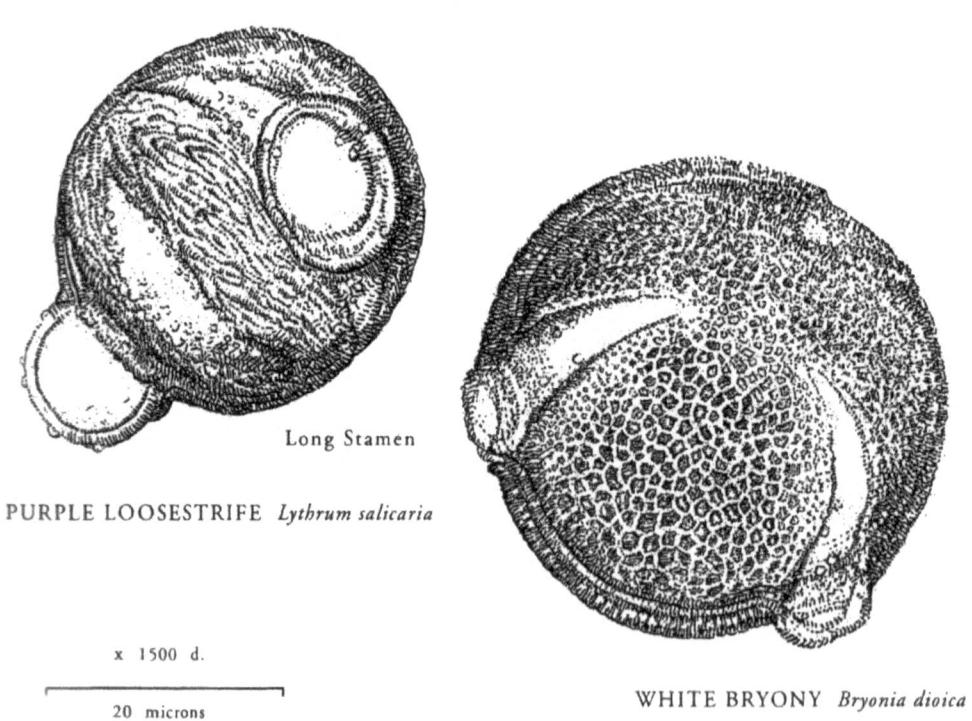

Long Stamen

PURPLE LOOSESTRIFE *Lythrum salicaria*

x 1500 d.

20 microns

WHITE BRYONY *Bryonia dioica*

LYTHRACEAE   CUCURRITACEAE

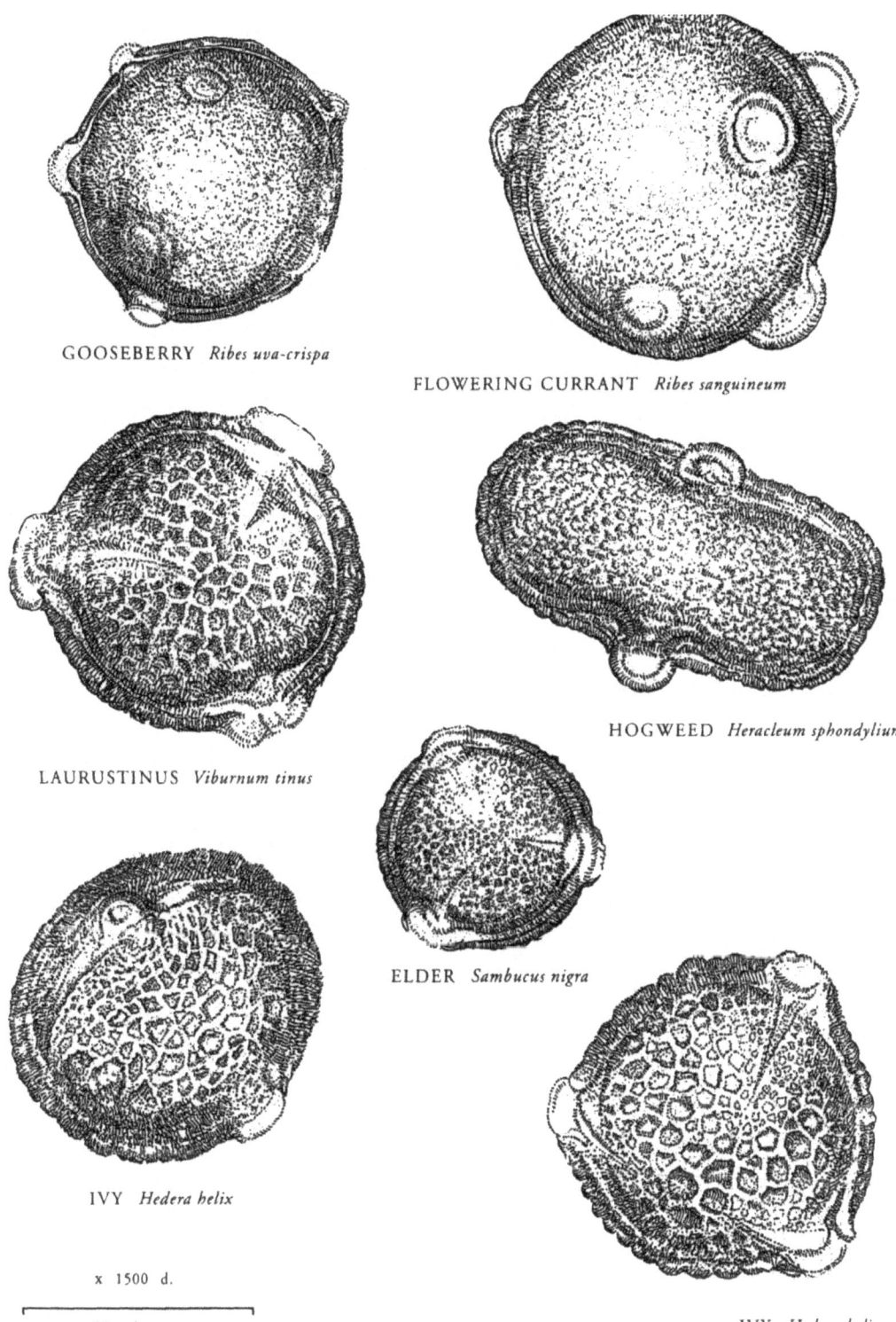

RIBESIACEAE UMBELLIFERAE CAPREIFOLIACEAE ARALIACEAE

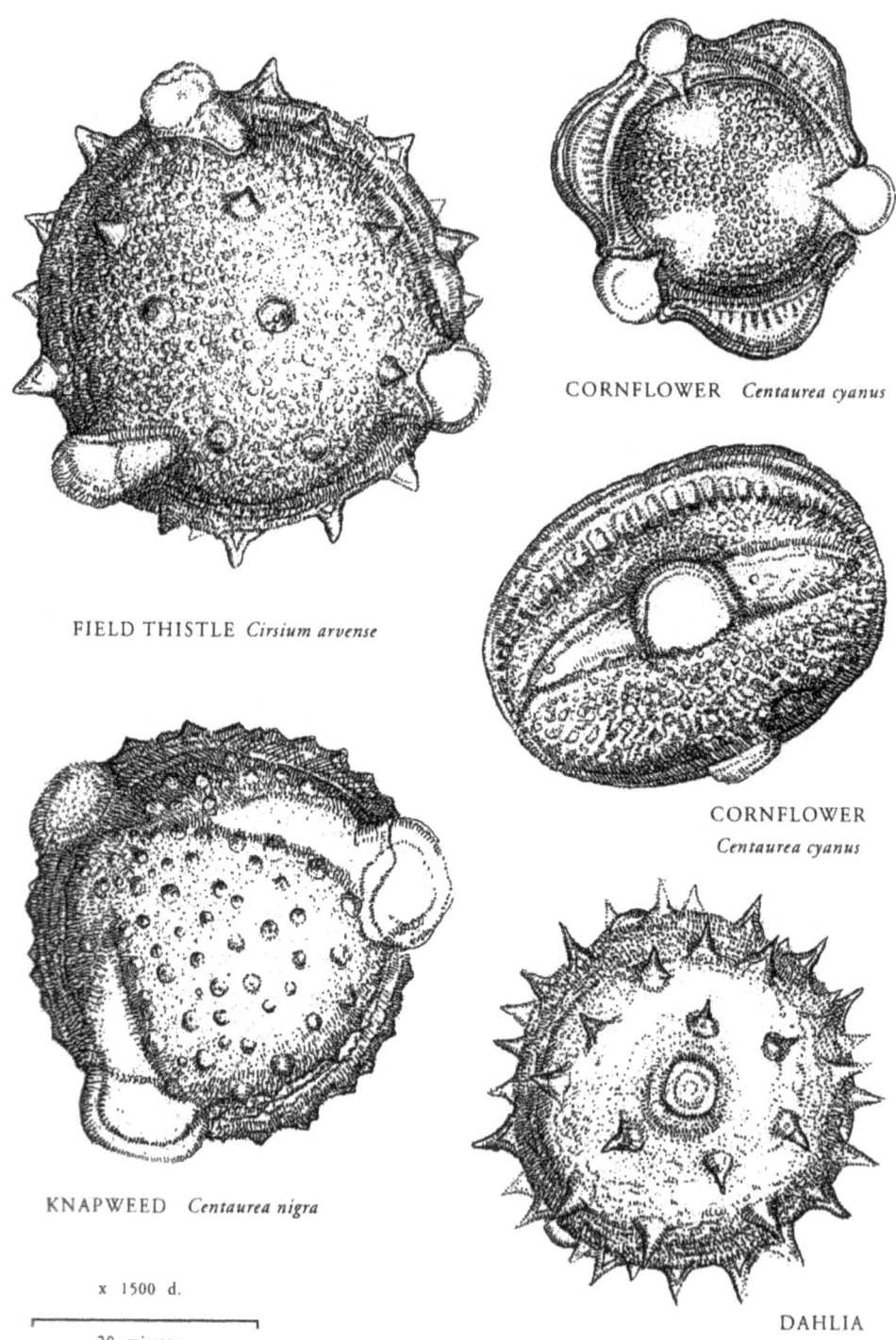

FIELD THISTLE *Cirsium arvense*

CORNFLOWER *Centaurea cyanus*

CORNFLOWER *Centaurea cyanus*

KNAPWEED *Centaurea nigra*

DAHLIA

x 1500 d.

20 microns

COMPOSITAE

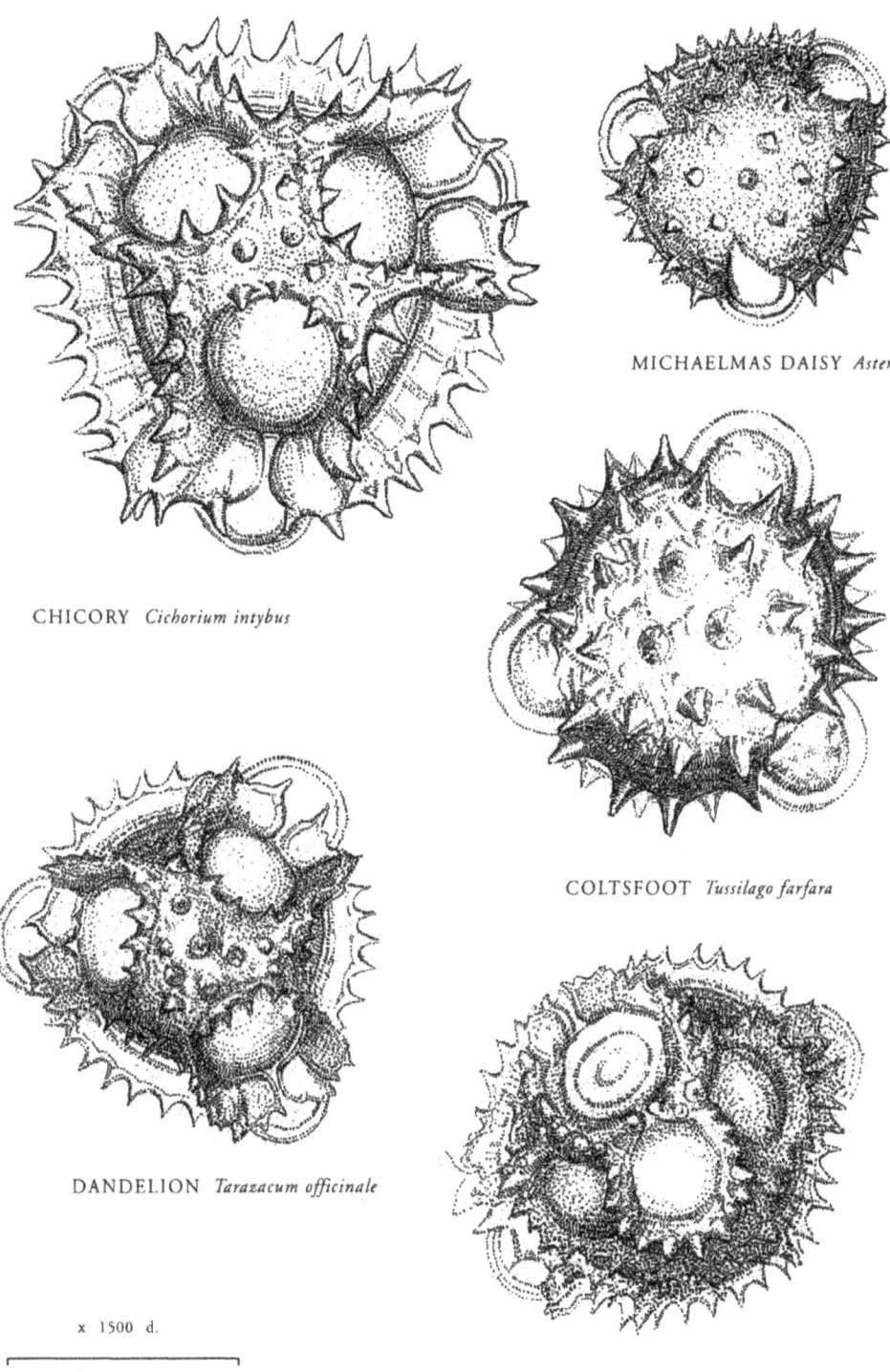

CHICORY *Cichorium intybus*

MICHAELMAS DAISY *Aster*

COLTSFOOT *Tussilago farfara*

DANDELION *Tarazacum officinale*

DANDELION *Tarazacum officinale*

x 1500 d.

20 microns

COMPOSITAE

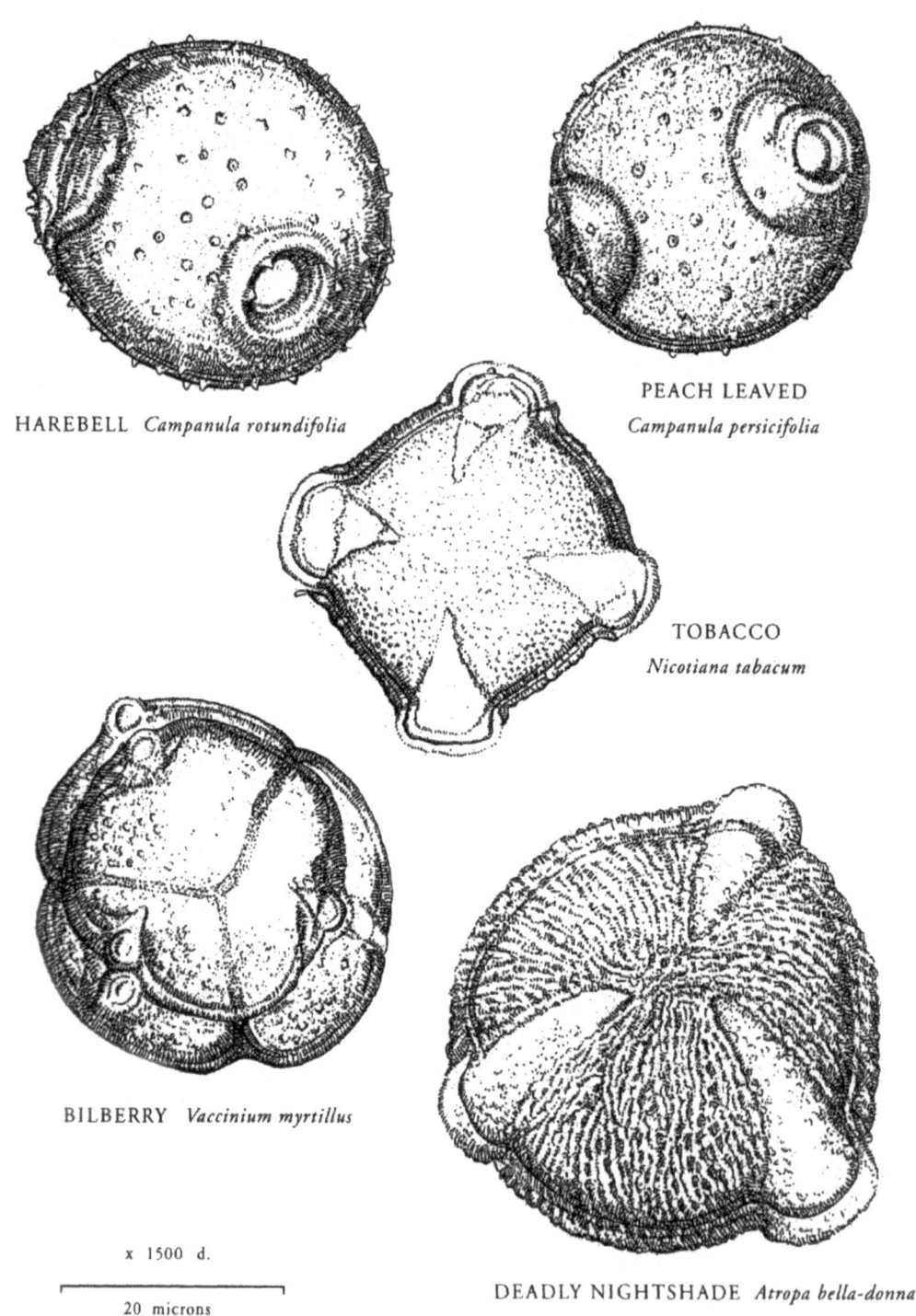

HAREBELL *Campanula rotundifolia*

PEACH LEAVED *Campanula persicifolia*

TOBACCO *Nicotiana tabacum*

BILBERRY *Vaccinium myrtillus*

x 1500 d.

20 microns

DEADLY NIGHTSHADE *Atropa bella-donna*

CAMPANULACEAE   SOLANACEAE   ERICACEAE

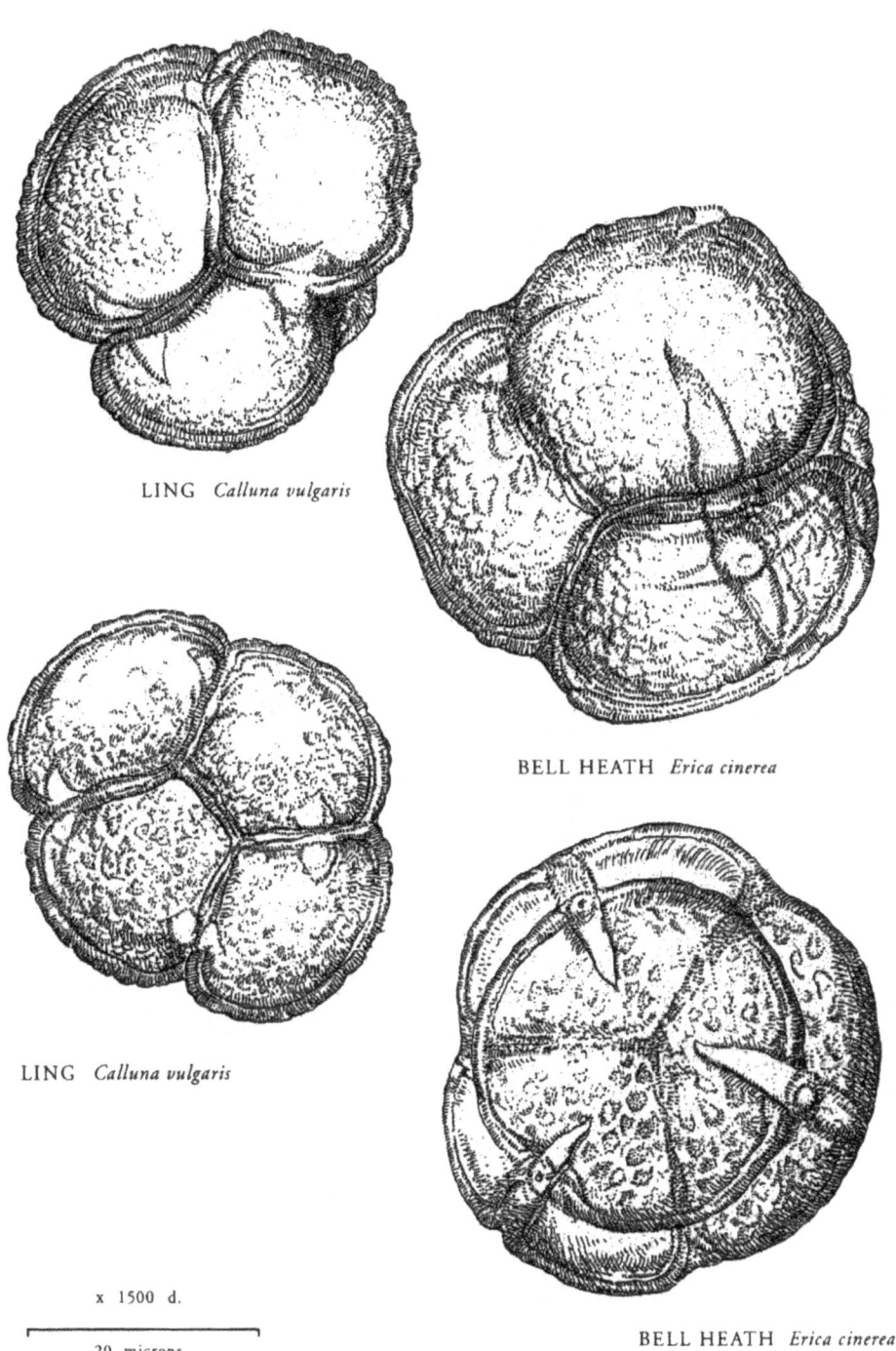

LING *Calluna vulgaris*

BELL HEATH *Erica cinerea*

LING *Calluna vulgaris*

BELL HEATH *Erica cinerea*

x 1500 d.

20 microns

ERICACEAE

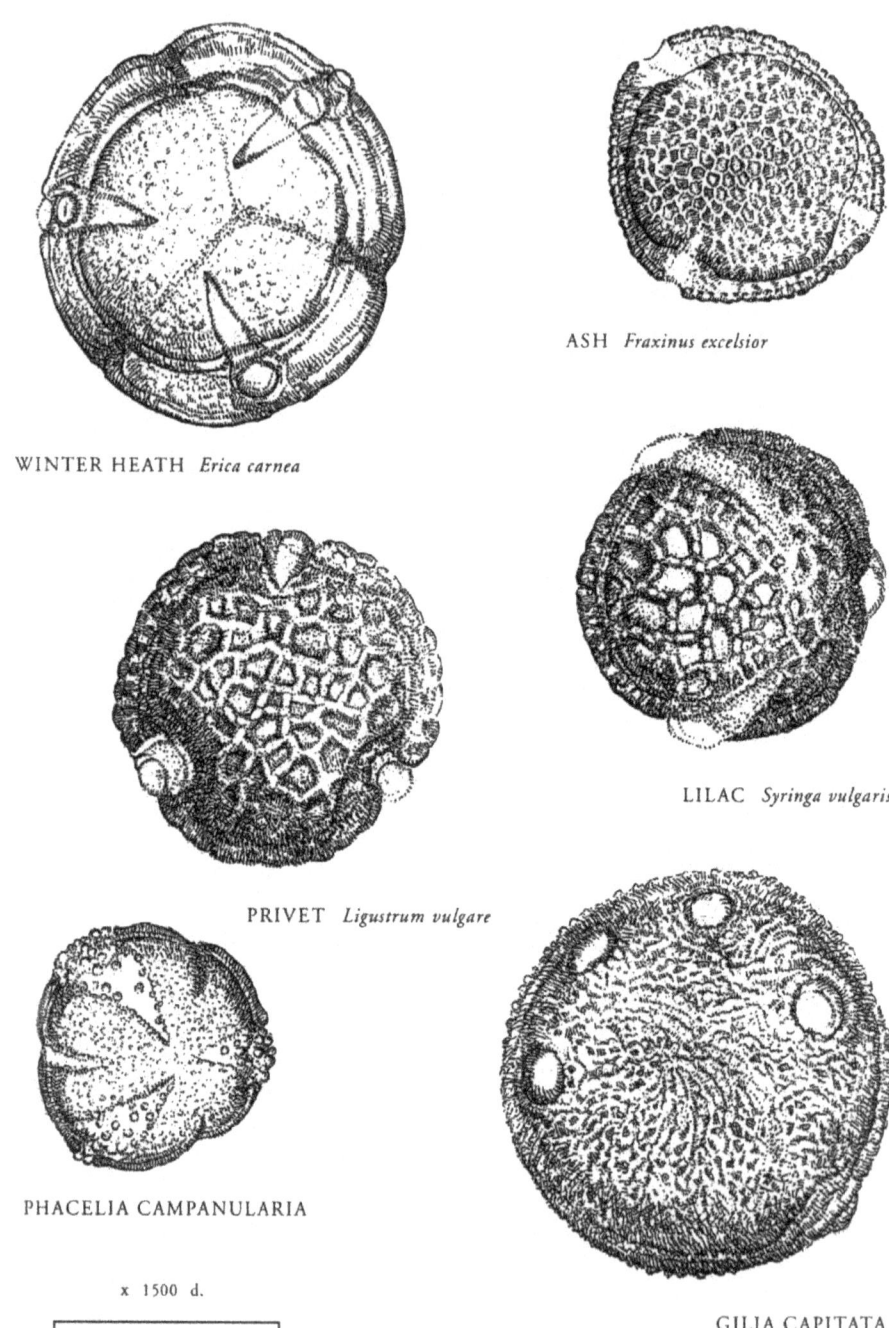

WINTER HEATH *Erica carnea*

ASH *Fraxinus excelsior*

PRIVET *Ligustrum vulgare*

LILAC *Syringa vulgaris*

PHACELIA CAMPANULARIA

GILIA CAPITATA

x 1500 d.

20 microns

ERICACEAE   OLEACEAE   POLEMONIACEAE   HYDROPHYLLACEAE

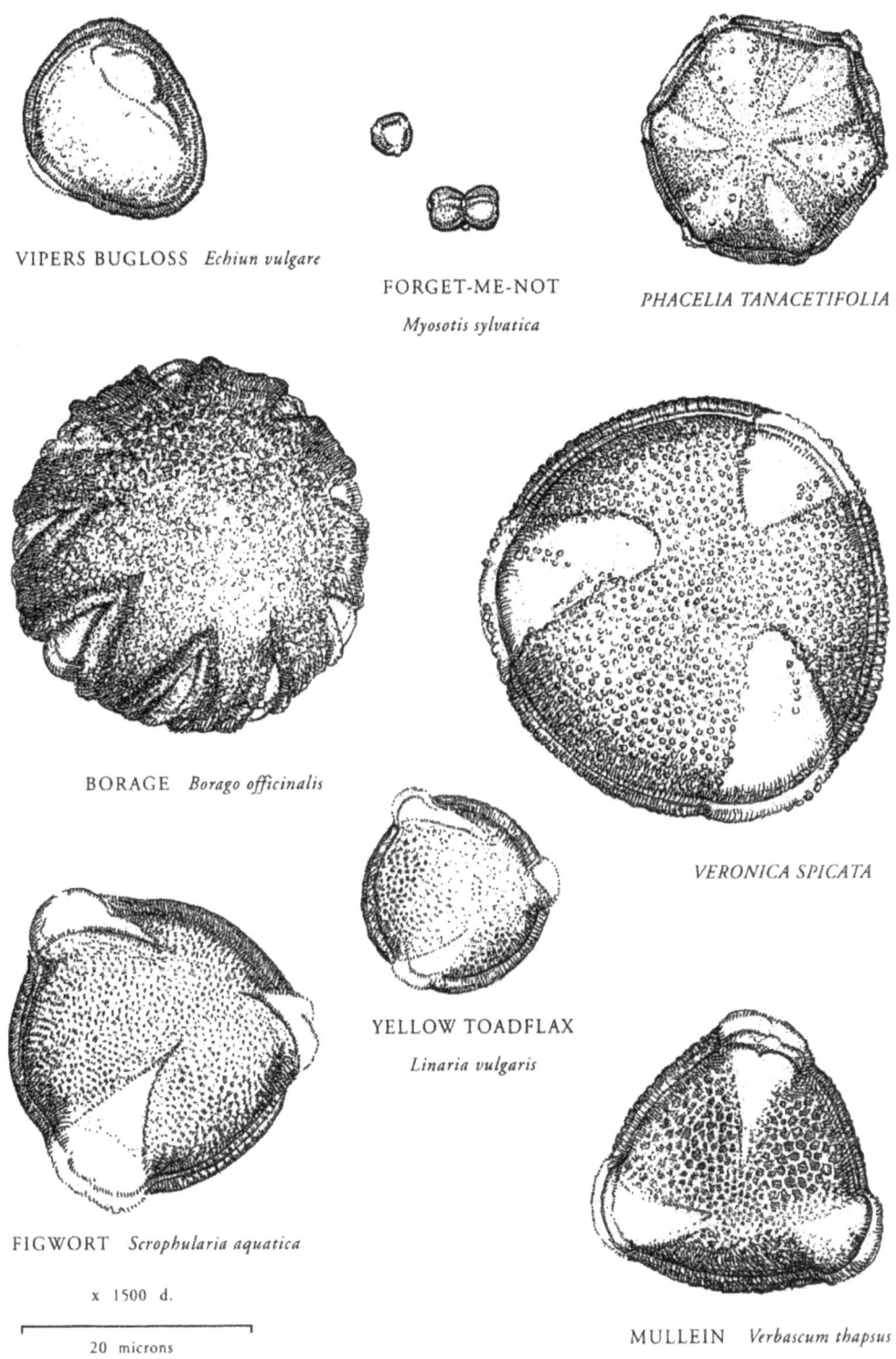

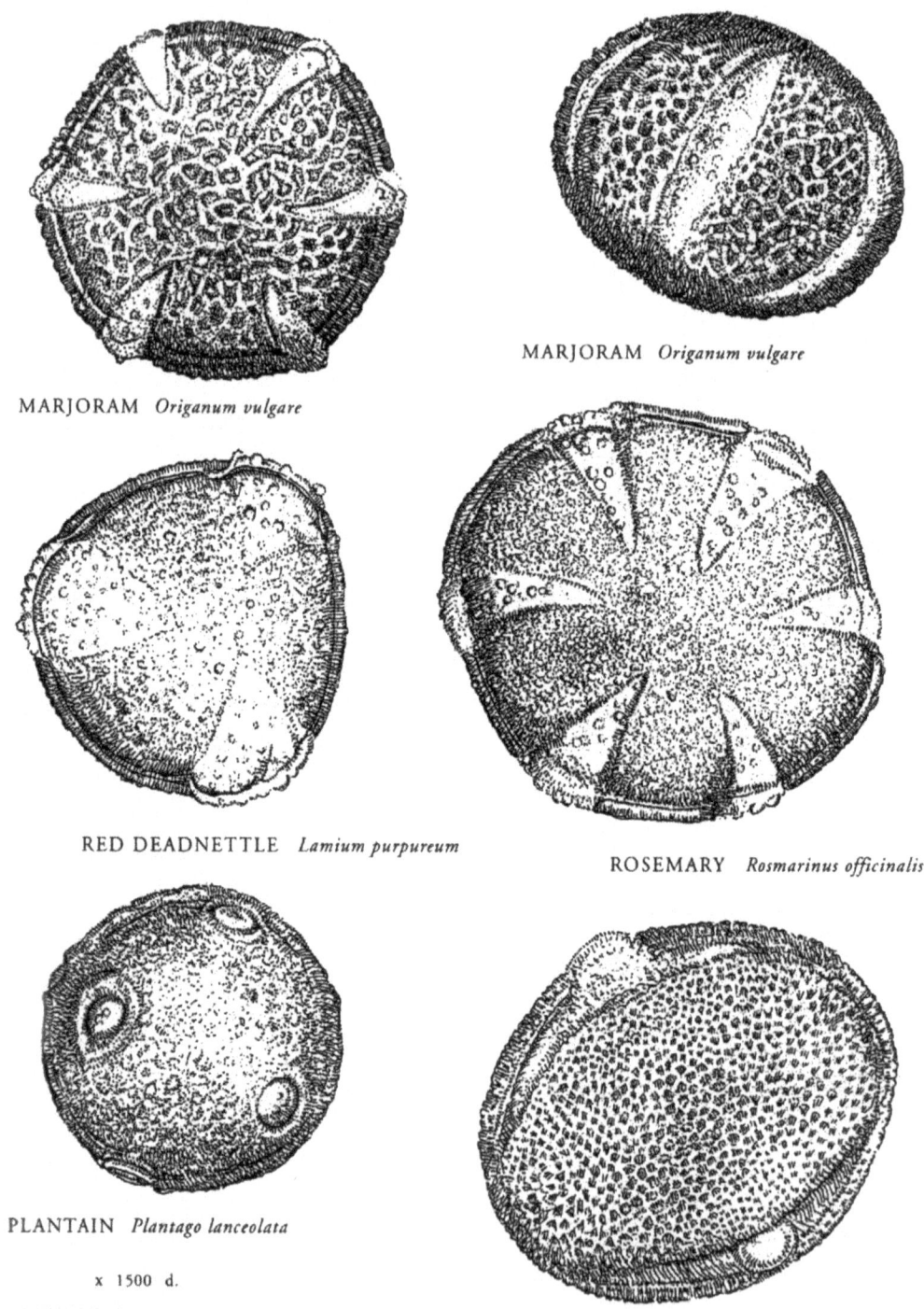

MARJORAM *Origanum vulgare*

MARJORAM *Origanum vulgare*

RED DEADNETTLE *Lamium purpureum*

ROSEMARY *Rosmarinus officinalis*

PLANTAIN *Plantago lanceolata*

x 1500 d.

20 microns

BUCKWHEAT *Fagopyrum esculentum*

LABIATAE   PLANTAGINACEAE   POLYGONACEAE

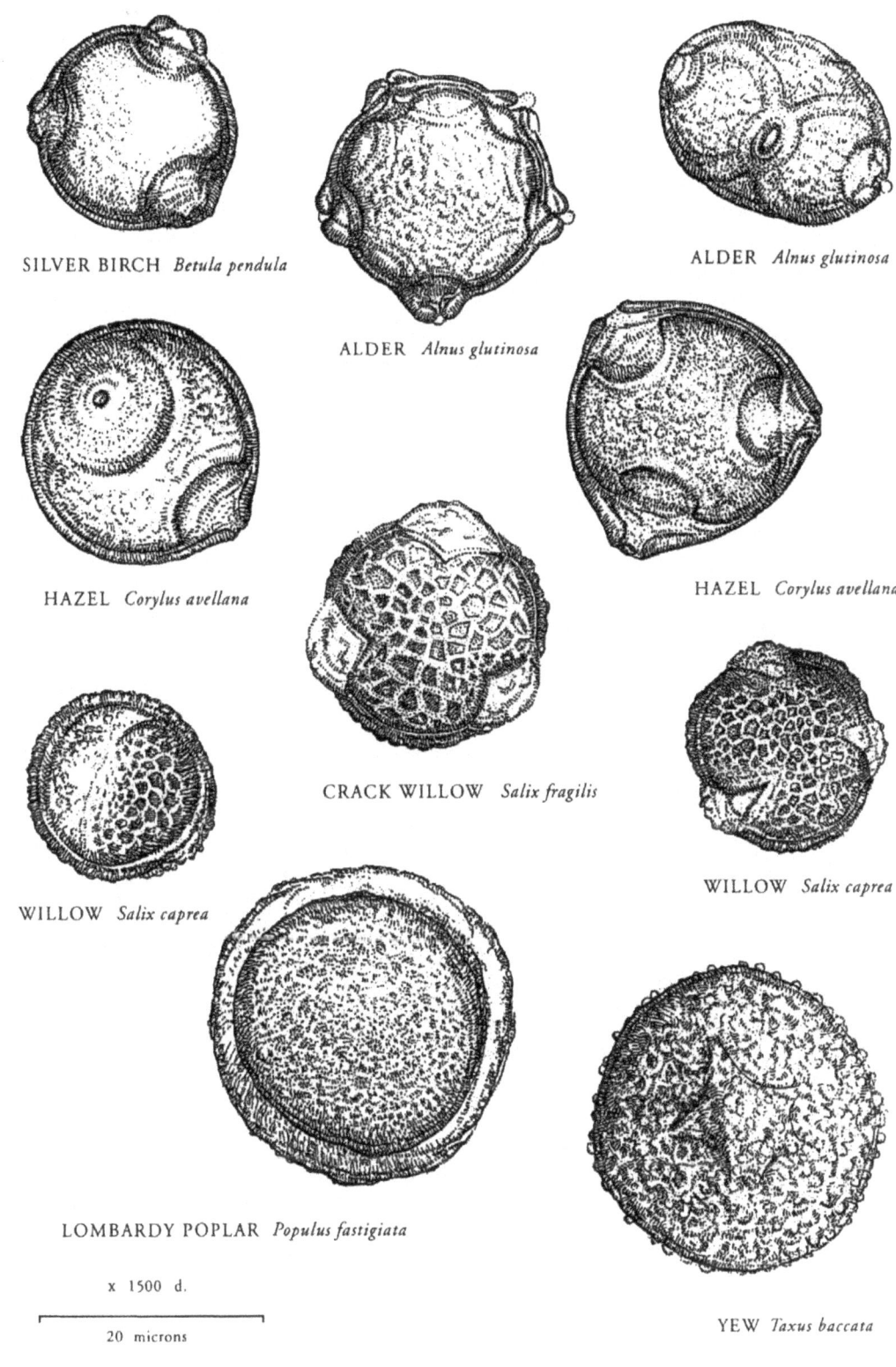

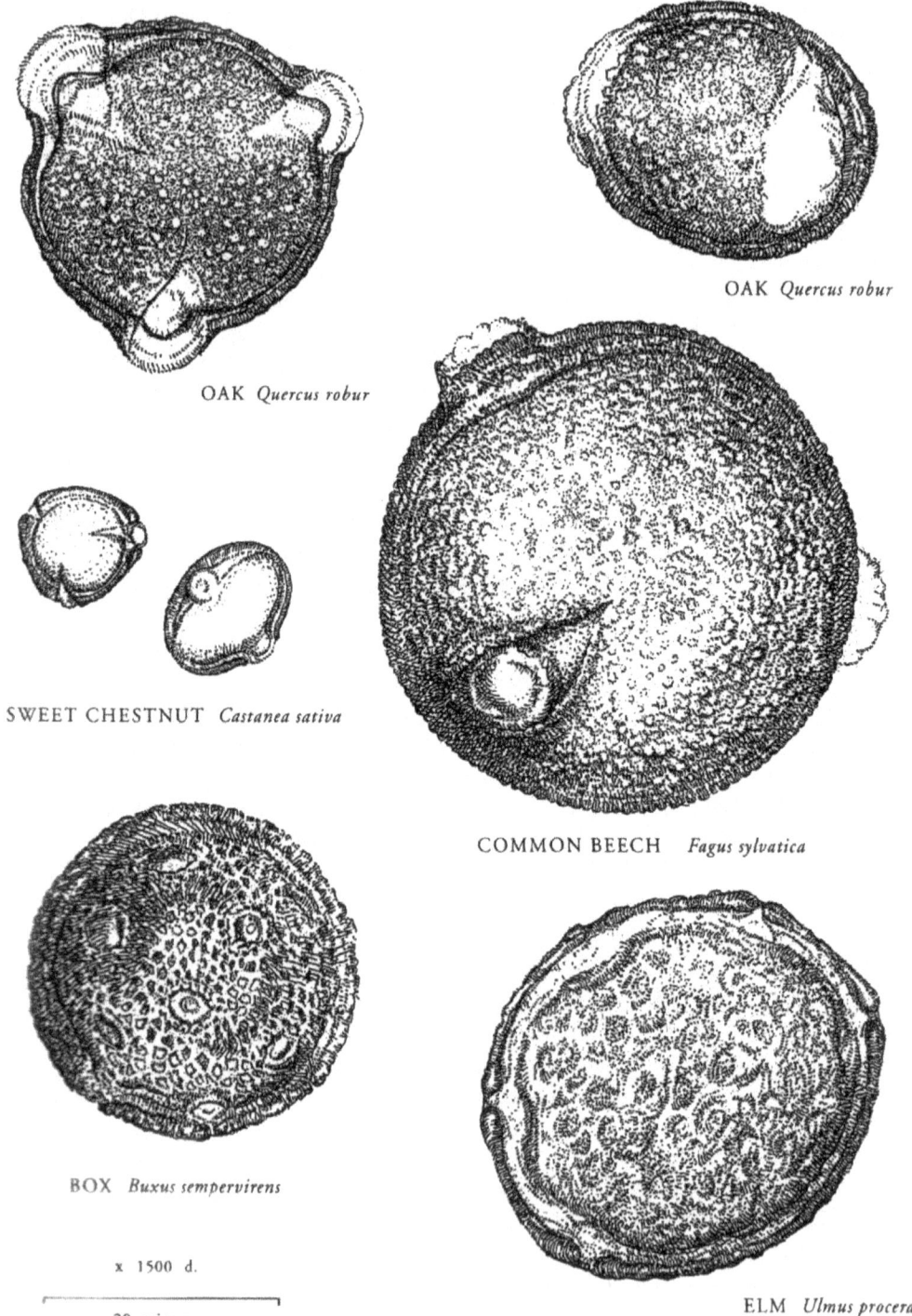

OAK *Quercus robur*

OAK *Quercus robur*

SWEET CHESTNUT *Castanea sativa*

COMMON BEECH *Fagus sylvatica*

BOX *Buxus sempervirens*

x 1500 d.

20 microns

ELM *Ulmus procera*

ULMACEAE   EUPHORBIACEAE   FAGACEAE

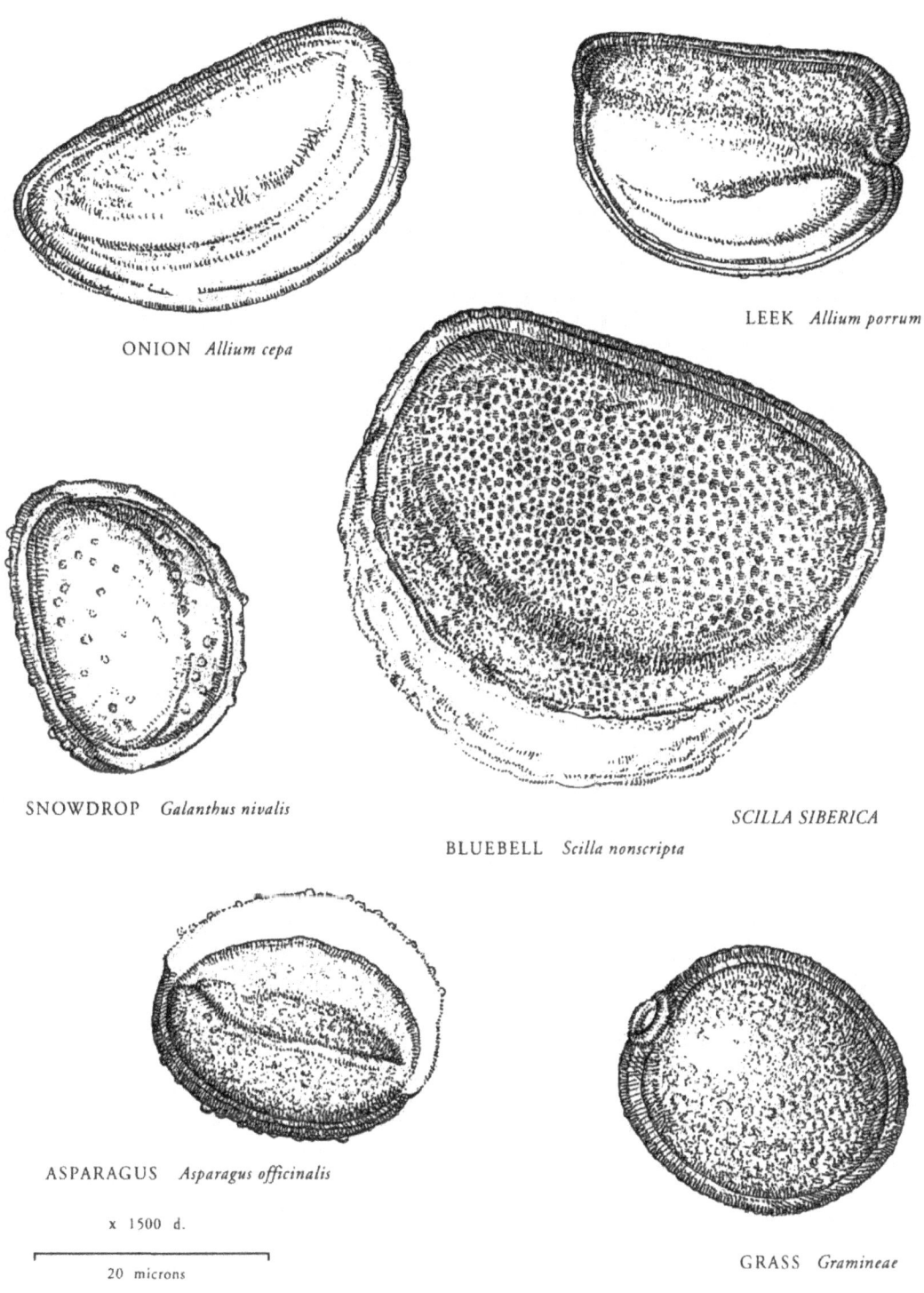

LILACEAE AMARYLLIDACEAE GRAMINEAE

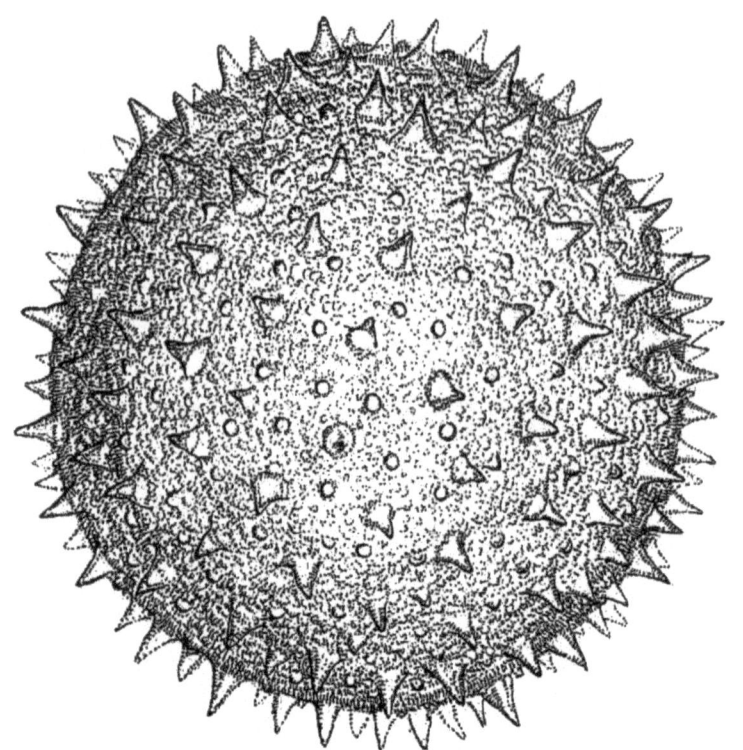

FORGET-ME-NOT *Myosotis sylvatica*

HOLLYHOCK *Althaea rosea*

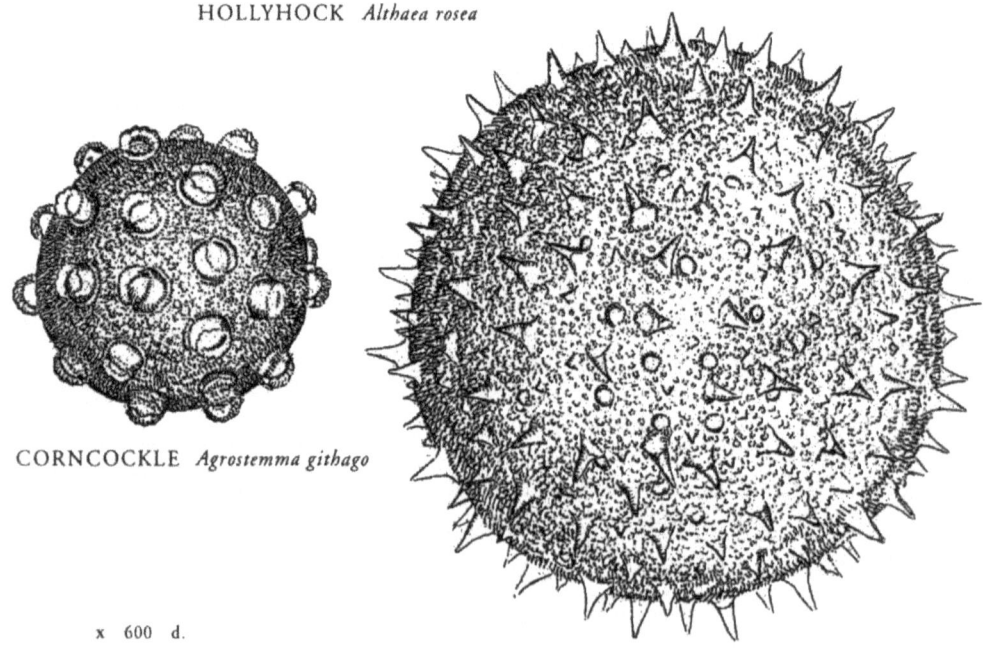

CORNCOCKLE *Agrostemma githago*

x 600 d.

50 microns

MALLOW *Malva sylvestris*

MALVACEAE   CARYOPHYLLACEAE   BORAGINACEAE

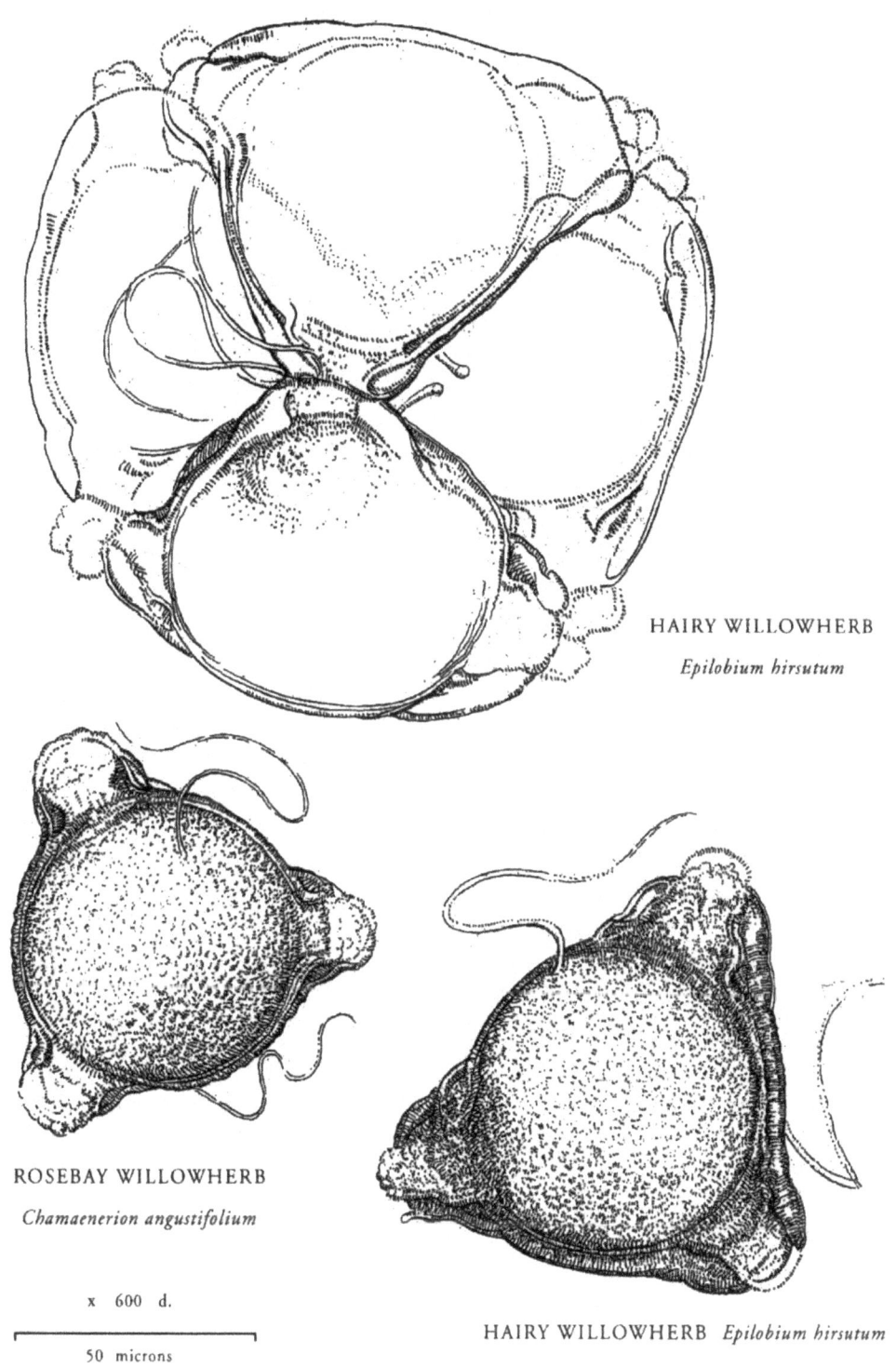

HAIRY WILLOWHERB
Epilobium hirsutum

ROSEBAY WILLOWHERB
Chamaenerion angustifolium

x 600 d.

50 microns

HAIRY WILLOWHERB *Epilobium hirsutum*

ONAGRACEAE

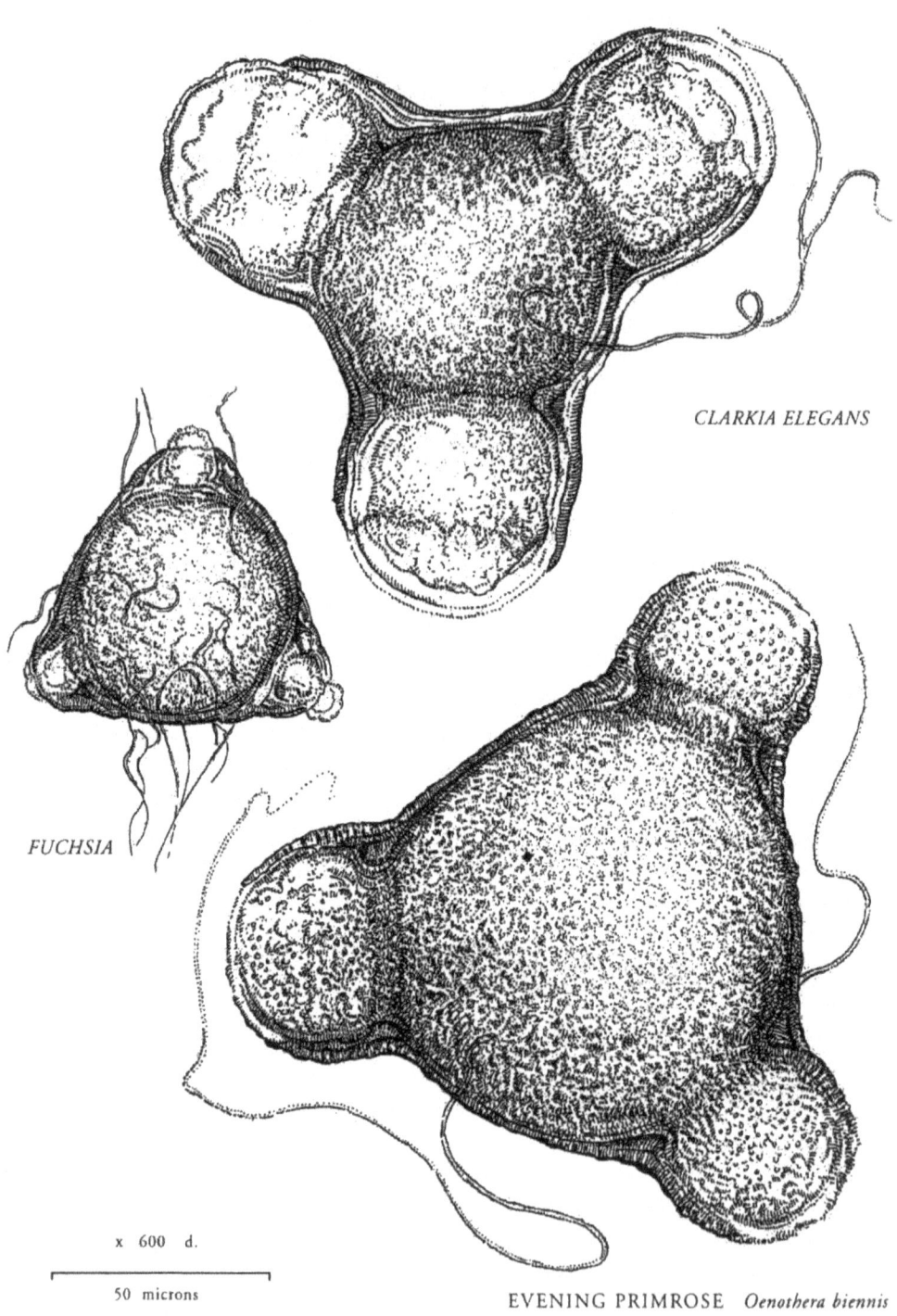

CLARKIA ELEGANS

FUCHSIA

EVENING PRIMROSE *Oenothera biennis*

x 600 d.

50 microns

ONAGRACEAE

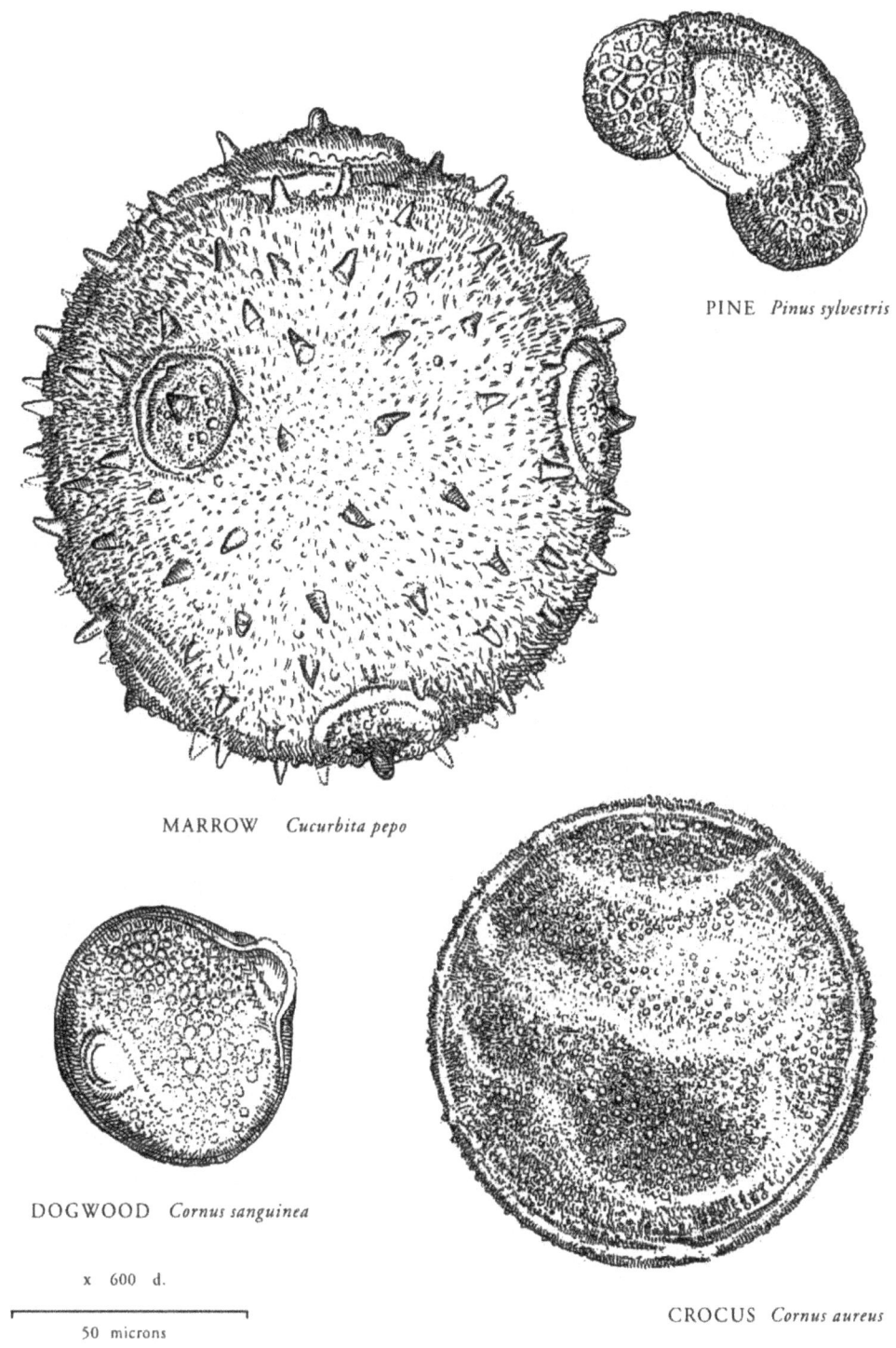

PINE *Pinus sylvestris*

MARROW *Cucurbita pepo*

DOGWOOD *Cornus sanguinea*

x 600 d.

50 microns

CROCUS *Cornus aureus*

CORNACEAE   IRIDACEAE   PINACEAE   CUCURBITACEAE

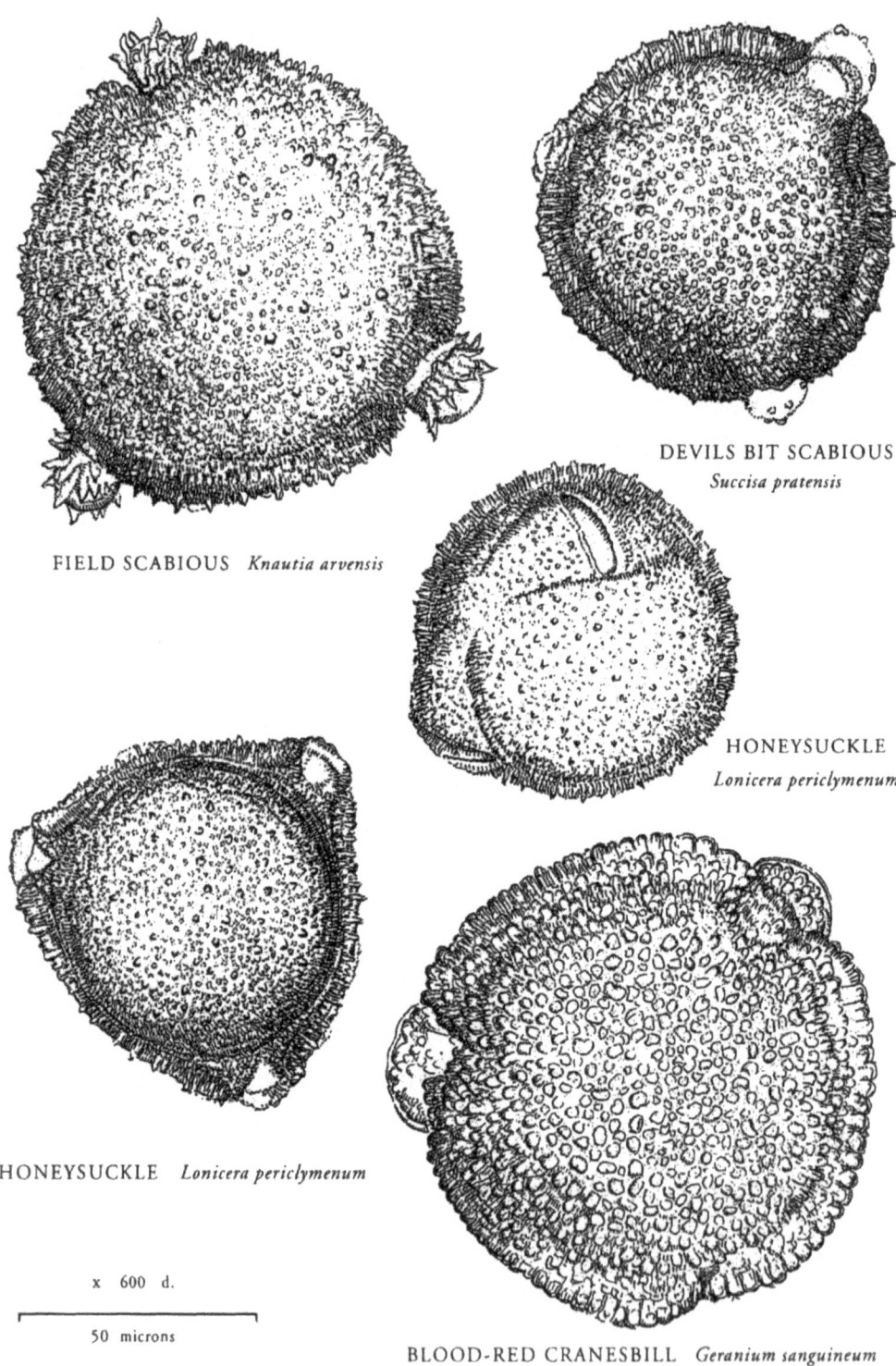

FIELD SCABIOUS *Knautia arvensis*

DEVILS BIT SCABIOUS *Succisa pratensis*

HONEYSUCKLE *Lonicera periclymenum*

HONEYSUCKLE *Lonicera periclymenum*

x 600 d.

50 microns

BLOOD-RED CRANESBILL *Geranium sanguineum*

DIPSACACEAE   GERANIACEAE   CAPRIFOLIACEAE

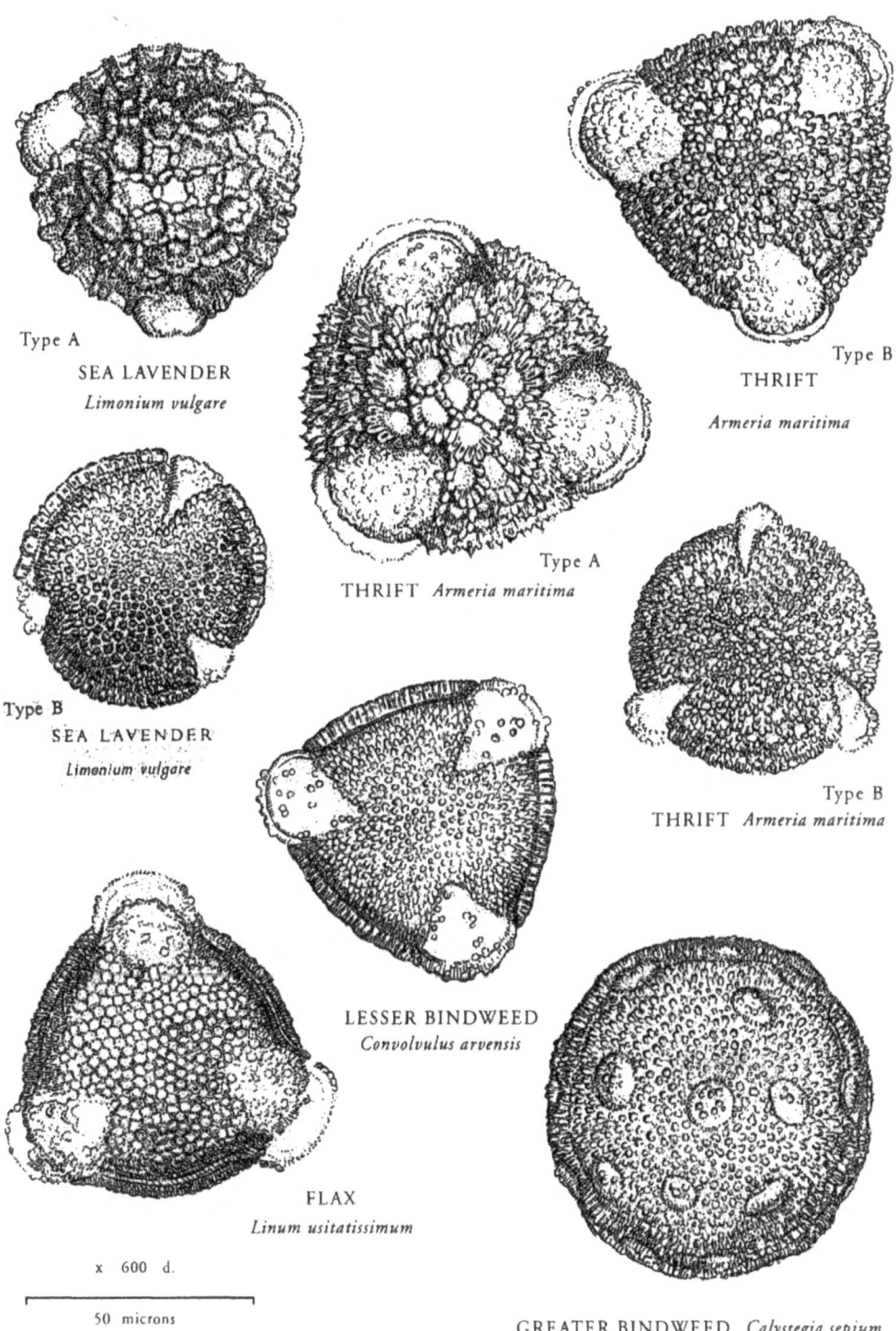

PLUMBAGINACEAE   LINACEAE   CONVOLVULACEAE

INDEX OF PLANTS

Acacia, False 6
Acer 4, 5
Aesculus 5
Agrostemma githago 25
Ailanthus altissima 5
Alder 22
Allium 24
Almond 10
Alnus glutinosa 22
Alsike Clover 7
Althaea rosea 25
Anemone nemorosa 1
Apple 11
Arabis alpina 2
Armeria maritima 30
Ash 19
Asparagus officinalis 24
Aster 16
Atropa belladonna 17
Aubrietia 2

Bean, Broad 8
Beech, Common 23
Berberis 1
Betula pendula 22
Bilberry 17
Bindweed 30
Birch 22
Birdsfoot Trefoil 6
Blackberry 12
Blackthora 9
Bluebell 24
Borago officinalis 20
Borage 20
Box 23
Brassica ooleracea 2
Broccoli 2
Broom 6
Bryonia dioica 13
Bryony White 13
Buckeye 5
Buckwheat 21
Bugloss, Viper's 20
Bunercup, Bulbous 1

Buxus sempervirens 23
Californian Poppy 2
Calluna vulgaris 18
Caltha palustris 1
Calystegia sepium 30
Campanula 17
Castanea sativa 23
Celandine, Lesser 1
Centaurea 15
Chamaenerion 26
Charlock 3
Cheiranthus 2
Cherry, Wild 10
Chestnut 5, 23
Chicory 16
Cichorium intybus 16
Cirsium arvense 15
Clarkia elegans 27
Clematis vitalba 1
Clover 7, 8
Coltsfoot 16
Convolvulus arvensis 30
Corncockle 25
Cornflower 15
Cornus sanguinea 28
Corylus avellana 22
Cranesbill 29
Crataegus monogyna 11
Crocus aureus 28
Cucurbita pepo 28
Currant, Flowering 14
Cydonia japonica 10

Dahlia 15
Dandelion 16
Deadly Nightshade 17
Deadnetde, Red 21
Dog Rose 9
Dogwood 28

Echium vulgare 20
Elder 14
Elm 23
Epilobium 26

Erica 18, 19
Eschscholtzia 2
Evening Primrose 27

Fagopyrum esculentum 21
Fagus sylatica 23
Figwort 20
Filipendula ulmaria 12
Flax 30
Forget-me-not 20, 25
Fraxinus excelsior 19
Fuchsia 27

Galanthus 24
Geranium sanguineum 29
Gilia capitata 19
Gooseberry 14
Gorse 6
Gramineae 24
Grass 24

Harebell 17
Hawthorn 11
Hazel 22
Hearh 18, 19
Hedera helix 14
Helianthemum 3
Heracleum sphondylium 14
Hogweed 14
Holly 5
Hollyhock 25
Honeysuckle 29

Ilex aquifolium 5
Ivy 14

Japanese Quince 10

Knapweed 15
Knautia arvensis 29

Lamium purpureum 21
Laurel 10
Laurustinus 14

INDEX OF PLANTS

Leek 24
Ligustrum vulgare 19
Lilac 19
Lime 4
Limonium vulgare 30
Linaria Vulgaris 20
Ling 18
Linum usitatissimum 30
Lonicera periclymenum 29
Loosestrife. Purple 13
Lotus corniculatus 6
Lucerne 8
Lupin 8
Lupinus polyphyllus 8
Lythrum salicaria 13

Mahonia aquifolia 1
Mallow 25
Malus pumila 11
Malva sylvestris 25
Maple 4, 5
Marigold, Marsh 1
Marjoram 21
Marrow, Vegetable 28
Meadowsweet 12
Medicago sativa 8
Melilotus 7
Michaelmas Daisy 16
Mignonette 3
Mountain Ash 12
Mullein 20
Mustard 3
Myosotis sylvatica 20, 25

Nicoriana tabacum 17
Nightshade, Deadly 17

Oak 23
Oenothera biennis 27
Old Man's Beard 1
Onion 24
Onobrychis viciifolia 8
Oriental Poppy 2
Origanum vulgare 21

Papaver 2
Parthenocissus tricuspidata ... 6
Pear 12
Phacelia 19, 20
Pine 28
Pinus sylvestris 28
Plantago 21
Plantain 21
Plum 9
Poplar, Lombardy 22
Poppy 2
Populus 22
Primrose, Evening 27
Privet 19
Prunus 9, 10
Purple Loosestrife 13
Pyrus 11, 12

Queen Anne's Thimble 19
Quercus robur 23
Quince, Japanese 10

Radish, Wild 3
Ranunculus 1
Raphanus raphanistrum 3
Raspberry 12
Red deadnettle 21
Reseda odorata 3
Ribes 14
Robinia pseudoacacia 6
Rockrose, Garden 3
Rosa canina 9
Rosemary 21
Rowan 12
Rubus 12

Sainfoin 8
Salix 22
Sambucus nigra 14
Sarothamnus scoparius 6
Scabious 29
Scilla 24
Scrophularia aquatica 20
Sea Lavender 30

Siberian Squill 24
Sinapis 3
Snowdrop 24
Sorbus aucuparia 12
Succisa pratensis 29
Sweet Chestnut 23
Sycamore 4
Syringa vulgaris 19

Taraxacum officinale 16
Toxus baccata 22
Thistle, Field 15
Thrift 30
Tilia 4
Toadflax, Yellow 20
Tobacco 17
Traveller's Joy 1
Tree of Heaven 5
Trifolium 7, 8
Tussilago farfara 16

Ulex 6
Ulmus procera 23

Vaccinium myrtillus 17
Verbascum thapsus 20
Veronica spicata 20
Viburnum tinus 14
Vicia faba 8
Viper's Bugloss 20
Virgiuia Creeper 6

Wallflower 2
Willow 22
Willowherb 26
Wood Anemone 1

Yew 22

www.ingramcontent.com/pod-product-compliance
Lightning Source LLC
Chambersburg PA
CBHW040544220526
45473CB00016B/3021